Apocalyptic Narratives

I0585814

Linking literature from the sociological study of the apocalyptic with the sociology and philosophy of science, *Apocalyptic Narratives* explores how the apocalyptic narrative frames and provides meaning to contemporary, secular and scientific crises focussing on nuclear war, general environmental crisis and climate change in both English- and German-speaking cultural contexts.

In particular, the book will use social identity and representation theories, the sociologies of risk and Lakatos' philosophy of science to trace how our cultural background and apocalyptic tradition shape our wider interpretation, communication and response to contemporary global crisis. The set of environmental and other challenges that the world is facing is often framed in terms of apocalyptic or existential crisis. Yet apocalyptic fears about the near future are nothing new. This book looks at the narrative connections between our current sense of crisis and the apocalyptic.

The book will be of interest to readers interested in environmental crisis and communication, the sociology and philosophy of science, and existential risk, but also to readers interested in the apocalyptic and its contemporary relevance.

Hauke Riesch is Senior Lecturer in Sociology at Brunel University London, researching science communication and the sociology and philosophy of science.

Routledge Studies in Science, Technology and Society

For the full list of books in the series: www.routledge.com/Routledge-Studies-in-Science-Technology-and-Society/book-series/SE0054

Apocalyptic Narratives

Science, Risk and Prophecy

Hauke Riesch

Routledge
Taylor & Francis Group

LONDON AND NEW YORK

First published 2021
by Routledge
2 Park Square, Milton Park, Abingdon, Oxon OX14 4RN

and by Routledge
605 Third Avenue, New York, NY 10158

Routledge is an imprint of the Taylor & Francis Group, an informa business

British Library Cataloguing-in-Publication Data
A catalogue record for this book is available from the British Library

Library of Congress Cataloging-in-Publication Data
Names: Riesch, Hauke, author.
Title: Apocalyptic narratives : science, risk and prophecy /
    Hauke Riesch.
Description: Abingdon, Oxon ; New York, NY : Routledge, 2021. |
    Series: Routledge studies in science, technology and society |
    Includes bibliographical references and index.
Identifiers: LCCN 2020054473 (print) | LCCN 2020054474
    (ebook) | ISBN 9780367275730 (hbk) |
    ISBN 9780429296680 (ebk)
Subjects: LCSH: Environmental disasters. | Environmental
    degradation. | Apocalypse in literature. | Risk communication. |
    Lakatos' model. | Science—Philosophy. | Science—Social aspects.
Classification: LCC GE140 .R545 2021 (print) | LCC GE140
    (ebook) | DDC 304.2/8—dc23
LC record available at https://lccn.loc.gov/2020054473
LC ebook record available at https://lccn.loc.gov/2020054474

ISBN: 978-0-367-27573-0 (hbk)
ISBN: 978-1-032-00608-6 (pbk)
ISBN: 978-0-429-29668-0 (ebk)

Typeset in Times New Roman
by Apex CoVantage, LLC

For Amaury

Contents

Acknowledgements

Like most other books, the origin story of this one is one of happenstance, coincidences and chance remarks. A few years ago, my colleagues and I in the Brunel sociology and media studies group decided to revamp our undergraduate degrees, introduce new modules and retire old ones. As part of this process, we all put forward suggestions for our dream modules. Since my specialisation lies in the sociology of science and risk, my contribution to this was to propose a module on the sociology of the environmental crisis. I thought this would be a wonderfully interesting proposition, but my colleague Meredith Jones suggested putting a bit more drama into the title, so I called it "Apocalypse! Crisis and Society". So, basically, I blame Meredith for this.

Once a module of that name was decided on, in order to justify it I put together a few lectures on the sociology of apocalypse and millennialism. Over time, as I started developing and then teaching the module, the two topics morphed into one in my mind; there are many parallels in how current environmental and technological crises and how apocalyptic and millennial visions are discussed. This book is the result of trying to put my thoughts down a bit more formally and develop an account of what links the apocalyptic imaginary with the problems our society faces today, and what the implications for that are in terms of our efforts of communicating and doing something about it.

The sociology of religion is a new area for me, and this caused some anxiety for me as I was writing this. I hope to have done justice to this field, but I'm sure at places some of the naivety of the newcomer will still shine through. Similarly, I have tried to be fair and even-handed in the treatment of the various religious beliefs here, given that in my own life I do not consider myself a religious person.

Acknowledgements need to go first to Meredith Jones, because her feedback on my module title started me off on this journey, but also for her help and support of this project. I would also like to say thanks to my other colleagues at Brunel, particularly Neil Stephens (who must be sick of hearing me talk about apocalypse by now but still helped me by reading through a few draft chapters), Neveen Abdalla, Isak Niehaus, Simon Weaver, Paul Moody, Steven Wainwright, and the Brunel Global Lives Research Centre. Outside of Brunel, Martin Bauer has been enormously supportive, as have my PhD supervisors Hasok Chang and Brian

Balmer. Also Kai Spiekermann, Grant Fisher, David Teira, Chiara Ambrosio, Daniel Helsing, Melanie Smallman, Matt Paskins, James Dolan, Jaron Harambam, William Griffin, Anne Cranny-Francis, David Spiegelhalter – all these poor people had to hear me go on about the apocalypse at various times and indulged me with good grace.

I have greatly benefited from presenting some of this work in the International Research Network for the Study of Science and Belief in Society conference in Birmingham, 2019. Again, I cannot stress enough how new the sociology of religion was for me as an area of research, and I was delighted to learn that my work was not only tolerated but actually encouraged. Thanks therefore to the new friends I made there, in particular to Fern Elsdon-Baker, James Riley, Thoko Kamwendo and Maria Rogińska.

The students who took my Apocalypse module have also been a fantastic source of comments, feedback and ideas, particularly so in the first year when I was still finding my feet with the material. I would like to say particular thanks to Hoodo Mohamed, Daniel Headlam and Jason Joshi. Farida Siddique even went as far as reading and commenting on one of my chapter drafts.

If my grant writing skill had been a little bit more persuasive, the previous few years of my life would have been spent researching science comedy instead, so some acknowledgement needs to go to the Arts and Humanities Research Council (AHRC) for not funding my project.

Finally, many, many, many thanks to my family, Werner, Eva, Robin and Malte Riesch, who have all in various amounts had to endure me talking about the apocalypse, and to my partner Bérangère Bacquey and our son Amaury for their support and patience.

Chapter 1

Introduction

The beginning of the end

I am writing this introduction in November 2020, a few days into a second national lockdown that the UK government has ordered in an attempt to curb the spread of a second wave of the coronavirus or COVID-19 pandemic – rather later than medical expert advice had advised, but better late than never. In the earlier phase of the pandemic, the first lockdown in April earlier this year appears, with hindsight, a time of solidarity; there was clapping for our National Health Service every Thursday and a feeling of togetherness, that we will overcome this obstacle. This had been especially heartening to see in a nation that had been bitterly divided by the process of exiting the European Union.

This time around, the lockdown has a different feel to it: Winter is coming, Christmas is ruined and there appears to be no end in sight to our new lives in relative isolation. We see our friends, families and colleagues much less often, our jobs are at risk, people – particularly my students cooped up in their halls – feel lonely and depressed, and there is the underlying fear for our health running in the background, and this is just for those of us who like me have so far been lucky enough not to have been personally affected by the virus. While there is still some solidarity, the lockdown rules are more relaxed, education institutions including my university are still open, but on the streets people seem to be trying their best to live their normal lives, and masks are worn only fairly infrequently even by those who take the disease seriously. There is an alarming international movement of people who believe that the virus is just a hoax, nothing that needs to be worried about other than as a conspiracy to keep people docile for the New World Order. This in turn has linked with the "QAnon" movement and grown into a larger contemporary conspiracy movement that links authoritarian politics, white supremacy and vaccine refusal in what some commentators have described as a culturally ascendant apocalyptic and millennial cult (Watt, 2020).

In short, this seems to be a perfect time to finish writing a book about the apocalypse. Allusions to the apocalyptic and the virus appear frequently, in pop culture memes that jokingly refer to zombie movies or post-apocalyptic landscapes, or newspaper cartoons that depict the virus as the fourth (or sometimes fifth) horseman of the apocalypse. Op-ed writers in the newspapers might warn about the apocalyptic as an interpretative frame or utilise it themselves (e.g. Scheller, 2020). The fact that all of us, worldwide, are facing the same issue adds to the pandemic appearing apocalyptic.

However, even before the virus hit, talk about worldwide crises had an often apocalyptic imagery attached to it: We would talk about climate change, catastrophic environmental crises (such as the collapse of insect populations), worldwide economic crisis or nuclear proliferation using an apocalyptic framing. The break-up of old political certainties, with populist and authoritarian regimes having come to power in several countries, with right-wing as well as religious terrorism on the rise and a growing new conspiracy culture surrounding QAnon that takes on older conspiracy narrative tropes of anti-Semitism, race wars and New World Order and links these with increasing success to current events (like the virus) – all this is happening with the climate crisis looming menacingly in the background. Commentators openly compare the current worldwide feeling of apocalyptic anxiety with that of the 1920s and '30s (e.g. Mason, 2016), and we all know where that led to.

As much as the feeling that the world is in grave peril is sincerely felt, it is also hardly a new one, and pervades contemporary political and public discussions (Wuthnow, 2010), in what Skrimshire calls the "politics of fear" (2008). Climate change has been widely recognised as a worrying issue since at least the 1990s, and before that the cold war held the attention of a whole generation that was convinced a civilisation ending nuclear war is either imminent or at least a real possibility. Before that, the world did actually flare up in two world wars separated by a worldwide economic crisis, which must have felt like the world was actually making a pretty good effort at ending, and not just a prediction. These fairly contemporary fears and crises are joined by a long line of apocalyptic and millennial/utopian expectations about the imminent end that stemmed from various religious denominations or humanist expectations of inevitable collapses of the current order like that of Marx or the Jacobins. As Skrimshire (2008) argued, current political crisis is conceptually linked to our apocalyptic, eschatological and utopian cultural heritage, and thus coloured accordingly. The invocation of the apocalyptic to current political crisis has been made by various academic discussants: Derrida (1984) discussed the politics of cold war nuclear fears in the mid-1980s through analogy with the apocalypse; Žižek (2010) likens the "terminal crisis" of capitalism to the end times. Morton (2013) describes the apocalyptic in terms of his "hyperobject" notion (objects that are too enormous to be perceived in full and include concepts such as climate change). What these discussions miss out on, for me, is a clearer sense of how the fascination with the apocalyptic translates into how we should think about current crisis and what makes it different from previous fears.

Because at the end (no pun intended!), almost every generation has had to live through either very justified apocalyptic fears or expectations or at least apocalyptic fears that feel very justified for those who believed in them – and yet the world has not ended. None of this is to say that current fears, certainly about climate change, are unsubstantiated, and clearly even if the world or civilisation survives it, this doesn't mean that there won't be plenty of suffering along the way. But if we have always been wrong before, is there anything that makes the current set of crises different (from a sociological rather than scientific point of view)?

This book represents my attempt to make some sense of how I should feel about my own apocalyptic fears – largely around climate change – by putting them into the context of the wider cultural heritage of apocalypse, and how the apocalypse as an interpretative narrative frame can push us into certain stances with regard to current fears. This includes not just how the apocalyptic narrative can enhance our anxieties but also how an apocalyptic narrative can make us complacent through the inductive argument that all predictions of the end have so far come to naught and therefore this one will too. I hope that by understanding the apocalyptic narrative frame a bit better, we can figure out better how to counter such arguments, how to use our experience and knowledge of the apocalyptic narrative to make a case for why certain apocalypses such as climate change should be taken much more seriously than they are.

It is here that I believe any analysis needs to take into account philosophies and sociologies of knowledge, how knowledge is arrived at and defended in the face of (disconfirming) evidence, and how all this relates to the underlying narratives we construct as a way of inserting meaning into our knowledge claims. Importantly, this needs to include scientific as well as non-scientific knowledge claims, since our reaction to the science of current crisis is being filtered through the lens of our wider cultural expectation of the apocalyptic. Overall, merely pointing to the science and its facts and evidences, as conclusive as this is for me personally, is not going to be enough to persuade anyone to take climate change (or any other apocalypse that one might be worried about) seriously, and it has been shown at length not to be enough as well (Gregory and Miller, 1998).

Apocalypses, millennia and conspiracies

The terminology surrounding academic literature on the apocalypse and apocalyptic thought can be confusing. It is not a clearly defined sociological term and tends to get used by scholars from different disciplines in subtly different ways. Wojcik (1997, p. 11) points out that biblical scholars tend to use it to describe the revelations and prophecies as limited to Jewish and Christian traditions delineated, for example, in the biblical books of Daniel and Revelation and the profuse historical array of their interpretations. Outside of biblical and theological studies, apocalypse has become a more general term and refers to "a sense of an ending, decline, societal crisis, and transformation" (Wojcik, pp. 11–12), and thus it does not rely on the divine or supernatural. It is this latter, more diffuse meaning that I will adopt, however I will also argue that the religious and the supernatural meaning of apocalypse has influenced and directed our wider apocalyptic visions to a large extent. It is in many ways part of the cultural baggage that we carry around with us and that we use to make sense of new and otherwise bewildering threats to our existence.

Another term often used within the literature is "millennialism". This is in some ways a more specific one, as it often connotes movements rather than mere beliefs, and again it is tied mostly to religious expectations of the end. Strictly speaking, the millennium here refers to a specific reading of scripture which roughly expects a specific sequence of events: Before the final battle between the forces of good and evil and God's final judgement, there will be a thousand years of peace and paradise – that is a millennium (Revelation 20:1–3, King James Version).

However, in many ways, though technically millennialism as a term is restricted to the various Christian interpretations of Revelation, the term is being applied more widely to sects, cults or movements that anticipate the imminent end of the world, where this end is anticipated because it will bring about a cleansing of the world, a period of peace and happiness and finally just punishment for sinners and the wicked. As such, millennialism can occur in other religions or movements that believe an imminent or currently ongoing period of social disruption will bring about either a period of earthly paradise or the final end followed with bliss for the righteous in the afterlife. In this vein, quite a variety of movements have been described as millennial, and these don't have to be religious in nature either, if they believe that the rupture and following period of bliss is a natural phenomenon (see Wessinger, 2011b). Thus they include communist revolutionaries (who, after all, believe in a new paradisiacal world order being brought about through revolutionary overthrow of the status quo), French revolutionaries, the German national socialists, ISIS and Islamic fundamentalism, UFO cults, New Agers who expect a new Age of Aquarius is imminent and many, many others.

Some scholars make a more rigid distinction between apocalypticism and millennialism; for example Robertson (2016) uses the term apocalyptic to refer to eschatologies (theological studies of the end) where "the outcome of the end time is total and destructive", whereas millennial narratives are eschatologies where "the world is not destroyed but transformed, and a better world is instigated" (p. 42). While having a clear terminology is obviously important, however, I will for the purposes of my argument here not employ such a rigid distinction. While I will recognise that apocalyptic and millennial have different connotations as to the destructiveness of the coming event as well as to the visions as to what will come after, there are considerable overlaps between the two, and there is rather a spectrum in both the destructiveness of the coming cataclysm and the desirability of the world afterwards that renders making such clear divisions difficult. But it is exactly the interpretative flexibility of these terms, alongside many of the other central ideas I will discuss (e.g. science, religion, risk) that makes them such powerful tools for the development of narratives and imaginaries of the near future.

Seen like this, millennialism and the apocalyptic are very fluid categories that include low profile and everyday movements as well as the big, destructive, revolutionary ones. After all, any movement advocating social change needs to make at least a little break with the past, and hope for an improved future. Thus some academic discussion on apocalypticism/millennialism includes popular cultural movements that would not necessarily or immediately be associated with either; for example Wojcik (1997) talks about the punk subcultural movement as displaying traits of the apocalyptic because of its nihilism and dystopian vision of where current society is headed. Next to punk, apocalyptic and millennial themes can be found in a variety of popular culture settings: For example the contributors in Walliss and Newport's (2014) collection examine apocalypticism in a variety of popular cultural contexts, ranging from the lyrics of Bob Dylan and heavy metal to literary genres like science fiction and anime. Within the sphere of political subcultures, some contemporary political projects such as Brexit or Trumpism have also been argued to have characteristics of a millennial movement and managed to mobilise evangelical voters through these associations (Knowles, 2018; Berry, 2020).

A type of social movement very much related to millennialism/apocalypticism, and one that I regret not being able to look at in more detail in this book, is that of the conspiracy theory. Conspiracy theories, as Cassam (2019) argues, are not just about there being nefarious conspiracies, because clearly people do in fact conspire very often. Instead, conspiracy theories "are first and foremost forms of political propaganda" (Cassam, 2019, p. 7), usually right wing, and this political association was also made prominent in Hofstadter's (1966) seminal essay on the "paranoid style of American politics". But there is also an epistemological strand to these. Harambam (2020) outlines academic literature on conspiracy theories: They have been linked to paranoid politics and societal danger but also to "bad science" (i.e. "flawed understandings of reality"). Conspiracy theory is described as explaining away all evidence to the contrary of their worldview as actually confirmatory to their idea that there is a vast conspiracy (that therefore also manipulates the available evidence); in Popper's (2005 [1959]) terms, then, they are unfalsifiable and therefore unscientific.

The link to politics and danger – as well as to millennialism – is probably well demonstrated by the rise of "prepper" culture, which encourages self-sufficiency and living apart from society in preparation of the imminent end (Garrett, 2020), but also the recent, more mainstream rise of the QAnon conspiracy theory during the build-up to Donald Trump's re-election campaign set against the backdrop of a global pandemic. A range of recent commentators discussing these movements have pointed to the link between apocalypticism and millennialism. For example, Jamie Doward, writing in the *Guardian*, introduces QAnon as "heavy on millennialism and the idea that a reckoning awaits the world, [and] the theory has found fertile ground in the American 'alt-right'" (Doward, 2020). The increasingly open association of QAnon with far-right, white supremacist and Nazi iconography (Lawrence, 2020) also indirectly points to a link, with Nazism noted by some prominent theorists in millennial studies as a prime example of a modern, secularised millennial movement (e.g. Barkun, 1974).

Michael Barkun has in more recent years himself developed an interest in studying conspiracy theories, which he describes as "apocalyptic visions" in the subtitle of his 2013 book. For Barkun, conspiracy theories and millennialism are linked, though not systematically so. Conspiracy theories, like millennialism, are often dualistic worldviews that divide the world into good and evil forces, where maybe a difference is that "conspiracy theories locate and describe evil, while millennialism explains the mechanism for its ultimate defeat" (Barkun, 2013, p. 10). Conspiracy theories are then a prominent element of the larger category of millennialism. But then again, making a clear distinction between these two concepts hinges on being able to settle on a clearer definition of what millennialism is in the first place.

Millennialism and conspiracy theory are themselves linked to wider discussion about how we know what we (think we) know. In that context, Barkun (2016) writes about conspiracy theories as "stigmatised knowledge", where its rejection by the authorities becomes "a sign that a belief must be true". Acknowledging that their beliefs lie on the fringe, the conspiracy theorists Robertson (2016) interviewed consciously described themselves as "conspiracy nutters". However, Barkun also notes that this appears to be changing, with the internet among other things blurring the boundaries between fringe and mainstream views. Harambam and Aupers (2015, 2017), in contrast to Robertson, found that conspiracy theorists they studied rejected the label as well as the stigmatised status of their worldviews, preferring instead to insist that their knowledge claims are based on science, reason and evidence.

In this, there is link between conspiracy theory as stigmatised knowledge and other, more overtly religious beliefs such as creationism, in that they too believe their view is supported by science and they use scientific legitimation strategies (setting up their own journals, conferences and learned societies) to argue their view. This then ties up to a wider argument I want to make in this book, that the apocalyptic knowledge claims we will encounter – from a formally religious point of view but also informally spiritual, conspiratorial and "properly" scientific – have similar legitimisation strategies, and all believe to be arguing from the side of "reason" and that therefore clear distinction between say secular and religious apocalyptic predictions is not so easily made. But I will of course expand on this point later on.

Narratives and meaning making

The way we think about the catastrophic near future is framed through our familiar cultural narratives which provide meaning to such frightening prospects. The cultural narratives of the apocalypse – struggles of good versus evil, prophets and messiahs who warn and save us, millennial utopias that rise out of the ashes, redemption and judgement – are all narratives that allow us to understand and make sense of contentious and frightening current developments of worldwide so-called existential risks such as climate change, environmental collapse or nuclear war. This will be one of my main arguments: By looking at some of the ways in which these secular apocalypses have been made sense of through wider cultural narratives, I intend to show how our responses to these risks are similarly infused (similar to Burgess' argument 2019 on the "risk" narrative in environmental discourse). This does not, however, mean that these contemporary scientific risks are merely current expressions of earlier apocalyptic fears. Clearly the science is there, and it points into genuinely worrying directions.

Instead I want to show how our secularised understanding of these existential crises forces us into the search for a wider meaning of it all. A completely nihilistic interpretation of, say, the climate crisis does nothing to help us. Apart from the fact that complete nihilism in the sense of a denial of meaning is quite impossible, since that itself is a way of making sense (Diken, 2009), taking action means identifying responsibility, planning for a post-apocalyptic future and finding solutions; all require that we understand what is going on, and we can only do that by attributing some ideas on what this means for us.

Thus my second main argument will be that our strategies for introducing meaning into our beliefs are not particularly different between science and other forms of apocalyptic knowledge making, and that this is as it should be. Science, like any other human activity, forms a social identity that responds to other beliefs through mechanisms of cultural cognition, evaluating evidence through the lens of its own background cultural and theoretical understandings and so on. I will elaborate this through first looking at the blurred boundaries between scientific and religious forms of making knowledge claims and then applying social representation theory, social identity theory, sociological theories of risk and Lakatos' philosophy of science to elaborate how apocalyptic narratives interact with our scientifically derived fears for the near future.

In this book I will talk about religions, religious movements, denominations, sects and cults. The dividing lines between these are not entirely clear, and particularly the usage of the terms "sect" and "cult" is frowned upon by some sociologists of religion as they are often used as pejorative terms. I have decided to use them, however, since sects and cults note a useful distinction between religions and religious movements such as denominations that indicate more mainstream acceptance. When I write about a movement being a sect or cult, therefore, this is to indicate that they are a more marginal grouping within the majority culture and is not meant to be pejorative or judgemental.

The end of the beginning

Chapter 2 will set the scene by outlining some of the literature on the differences and purported conflicts between science and religion, arguing that the boundaries between the two are blurred and that neither concept ever received a particularly useful definition. Instead I will suggest that they should be looked at as different epistemic communities that can be looked at through their differing social identities whose knowledge claims can and do influence each other and which share the fundamentally similar aim of making sense of the world.

Chapter 3 will provide a very short outline and history of the apocalyptic and its narrative tropes, ending with an overview of some of the main sociological attempts to explain what makes the apocalyptic such a persuasive and recurring proposition: Why do people so often tend to believe in a near-future apocalypse?

Chapter 4 will look at the apocalypse through the social representations of the various imaginaries of what the end will be, how severe it will be, what will happen and whose responsibility it is. This will link the contemporary scientific "existential risks" (and the "existential risk" concept itself) with scriptural and contemporary fictional accounts of the end, using Moscovici's theory of social representation.

Chapter 5 will focus on the failure of apocalyptic prophecies, linking Festinger's social psychological account of failed prophecy in apocalyptic movements to Lakatos' philosophy of science, arguing that they both attempt to provide accounts of how epistemic groups deal with disconfirming evidence and pointing to the parallels between them. This, I will argue, will give us a conceptual tool through which we can relate how both science and religious movements make knowledge claims.

Chapter 6 will present the concept of risk as a modern explanatory tool for explaining and making sense of near-future catastrophic events. I will outline risk society, the cultural theory of risk and social identity approaches to risk and how they can be used to make sense of our interpretations of the apocalyptic (through the running example of the coronavirus crisis).

Chapter 7 will look at the relations between the millennium and utopia, arguing that utopia in contemporary discourses on technological risks functions as the millennial myth that allows us to insert hope into the future, but that it can also be fundamentally divisive: A utopia is always only a utopia for those who dream it up.

Chapters 8–10 will apply these apocalyptic narratives, epistemologies and sociologies to three contemporary and widely discussed technological crises: Nuclear war, environmentalism and climate change. I will draw mainly from English- and German-language accounts in an attempt to widen the discussion beyond the usual Anglophone focus, with Germany having been chosen because of my familiarity with the country and the language. But Germany is also an interesting contrast in these debates because of the country's history in nuclear weapons research and its unique split position in the cold war, as well as its contemporary ambition and self-perception as a leader on environmental issues. These three themes will be introduced through pointing out their narrative connections with spiritual and religious discourses through their millennial imaginaries, prophetic figures and conceptions of nature.

Chapter 2

Making sense of how we make sense of the world

Introduction

A clear temptation in analysing apocalyptic narratives is to divide discourses into religious apocalypses, such as those of various apocalyptic cults, movements and religious traditions from the secular, or scientific ones, such as those that surround current fears about the environment or meteor impacts. This is done in the expectation that secular/scientific and religious narratives (and the beliefs that underpin them) require different types of explanation. While people worried about environmental apocalypse can be analysed as a millennial movement (Globus and Taylor, 2011), the difference is still that their fears are based on reason and science rather than the religious motives that drive "traditional" millennial movements (Wojcik, 1997; Daniels, 1999).

One of the main arguments I will be developing is that we can interpret our contemporary, largely scientific responses to worldwide crisis as partly derived from and influenced by older and more established apocalyptic narratives, and that this colours our responses to current, "secular" crises. The way we make sense of new and unfamiliar concepts and events is by anchoring them conceptually to our familiar and already digested body of social knowledge and then objectifying (or reifying) them as the new concepts become absorbed into our now-expanded body of social knowledge. I will present this argument more fully in Chapter 4 using Serge Moscovici's (2001) theory of social representation.

The familiar body of social knowledge, in the case of the apocalyptic and the end-of-the-world narrative and interpretative repertoires we use in order to make sense of contemporary crisis, is mostly religious rather than scientific in character simply because most of our cultural heritage is derived from pre-scientific times. This might present a clash of values in a contemporary world where the scientific and the religious tend to be seen as incompatible knowledge systems, where, moreover, one of them is usually presented as rational and epistemically warranted and the other is seen as an outdated, irrational or superseded mode of thinking that directly clashes with the first – or at least that is the caricature.

An instinctual reaction to my thesis may then be that to interpret scientific understanding through the lens of religious conceptions of the apocalyptic is to diminish or play down the empirically derived truth claims of the sciences. This is not my intention. However, I believe that knowledge claims need to be evaluated as a whole and, from a sociologists' perspective, neutrally or agnostically. In order to be agnostic towards knowledge claims, I believe my thesis needs to be presented as a sociology of knowledge, or belief, rather than a sociology of science (or religion). To borrow the words of Frank Kermode (1967), my aim is to make sense of how we make sense of the world.

In order to do this, I would in this chapter like to explore how a sociology of knowledge that combines religious and scientific knowledge can look: First by having a look at traditional and contemporary scholarship on the distinctions between the two forms of knowledge claims, and then by exploring how the sociological (and philosophical) scholarship of the two can be consolidated.

Science and religion

The perception that science is incompatible with religion has become a particularly fervently held view. Proponents of the movement sometimes labelled "new atheism", such as Sam Harris, Daniel Dennett, A. C. Grayling and Richard Dawkins (Cimino and Smith, 2011), have prominently and publicly argued that belief is incompatible with being a scientist (and thus, as scientists or science supporters, have come to identify themselves as atheists). The philosopher of science John Worral has argued that scientists who are religious either don't think things through properly, fail to be properly scientific or hold an "untenable" position that science and religion cover different domains (Worral, 2004). But even outside the more polemical pronouncements of the new atheists, a general belief that science has contributed as a major factor to the secularisation of (Western) society (as for example argued by Habermas, cited in Rogińska, 2019) is firmly entrenched.

In a wider academic setting, however, a large body of scholarship has been exploring their interactions and interrelations (De Cruz, 2019). This academic field on the relation between science and religion has its own journals (such as *Zygon*), academic conferences and societies, and it features contributors from the social sciences, philosophy and theology and natural scientists interested in the area, usually due to their own religious beliefs. Among this body of scholarship, the view that there is an inherent conflict between science and religion is mostly discredited (Harrison, 2010).

First, we can point to historical studies on the influence that religious institutions had on the preservation of religious knowledge through the European Dark Ages, that it was the religious establishment that drove the development of universities (Lindberg, 2010), or that almost all of the main figures from the scientific revolution, such as Isaac Newton (Austin, 1970), were deeply religious themselves. This on its own is not a conclusive argument, and one could counter this by arguing (using Worral's distinctions) that the early scientific pioneers didn't really think this through fully and/or maintained a religious belief in spite of its incompatibility with science because science itself hadn't fully matured at that stage, or that, finally, people then had an untenable view that the two domains could be held apart.

However, one can also argue that whatever the situation was then, things are now very different. Science has matured and discovered fundamental truths about the universe which stand in direct conflict with religious teaching, and therefore, while they may have been compatible before, they are not now (Stenmark, 2010). This of course dismisses a large number of contemporary religious scientists who reject this conflict view as either irrational, disingenuous or bad at science. While research shows that scientists on the whole are less religious than the wider population (e.g. Ecklund and Scheitle, 2007), this often rests on Western, mostly Anglophone societies, which may instead indicate that it is the specific cultural connections that drive how scientists interact with religion rather than there being an inherent conflict (see e.g. Rogińska, 2019, on scientists in Eastern Europe and Geraci, 2018, on scientists in South India).

Theorising the relationship between science and religion usually departs from Ian Barbour's classic work (2000; Berg, 2002). Barbour divides the ways in which the two can interact into four types of relationship. First, the "conflict" model sees the two as both incompatible and in direct conflict with each other. Arguments like those of the new atheists, for example, contend that they make conflicting truth or knowledge claims and that of those two, at most one can be true. Crudely presented like this, that is a logical position to take, as two competing claims cannot both be true, however it also rests on a very specific and not universally shared assumption on what types of knowledge claims the two are making. Second, an "independence" model would then posit that science and religion can be distinguished by the questions, domains and methodologies they employ, and that because these three all cover different ground, religion and science can be kept, in Barbour's words, "in watertight compartments" (2000, p. 17). This is the position taken for example by the "Non-overlapping Magisteria" (NOMA) concept formulated by S. J. Gould (1997; see also Moritz, 2009). Third, "dialogue" interactions emphasise the similarities and parallels, rather than the differences, in the questions, domains and methodologies of the two. Fourth, "integration" seeks a synthesis where both science and religion contribute to an overall inclusive view of the world.

As a scientist who was motivated to analyse the relationships between science and religion due to his own religious backgrounds and beliefs, unsurprisingly Barbour's own sympathies lie with the dialogue and integration models. This debate, not just Barbour's contribution, quite often also takes a very Christian view on the nature of religion. Regardless of where one stands, however, seeking to classify the relationship into qualitatively different types of possible viewpoints is a sensible point of departure. While later scholars have refined his model or even disagreed and posited their own, Barbour's analysis has been widely influential.

Underpinning these arguments about the compatibility and interrelations of science and religion, however, are wider conflicts around what exactly religion and science are; how exactly they interact and where the boundaries between them lies clearly depends on how we conceptualise them. One of the reasons little agreement is reached is that in both cases the boundaries are fluid, disputed and, in most practical applications, usually left undefined. We don't really need a clear vision of what religion is in order to believe in, say, redemption through Christ any more than we need to know the exact definition of science in order to do physics (Riesch, 2008).

Accounts of what science and religion are have been changing over the centuries, and in a way, even talking about conflict between them in previous centuries makes little sense. To explain this, Harrison's (2015) book on the evolution of the terms and their relations uses a territorial metaphor: "If a historian were to contend that he or she had discovered evidence of a hitherto unknown war that had broken out in the year 1600 between Israel and Egypt, this claim would be treated with some scepticism" (p. 1). This is because although the territory on which the modern nations of Israel and Egypt now stand existed at the time, it was carved up and organised differently (both being part of the larger entity of the Ottoman Empire). Similarly, while things like belief, faith and reason have always been part of our social life, carving them up into religious and scientific parts is a very modern intervention. Arguments that science and religion have not always been in conflict with each other are therefore not just not true: Even talking about pre-modern religion and science doesn't make much sense, or at least we would need to refer to social activities that are very different to what we would now understand as science or religion.

That the contemporary concept of science has been developing (and still is in the process of developing) over the centuries is probably not a contentious issue. The received story is that science has developed into its recognisable modern shape, and as a separate social practice, at the start of the modern era during the scientific revolution. Refinements and developments in what is meant by science, how science arrives at its conclusions and even why we do science are still going on, but essentially science, in its modern form at least, is a renaissance invention. Of course, the picture is more complicated than that (Shapin, 1996): The practice of trying to systematically make sense of the world is much older than the renaissance, and therefore history of science quite sensibly also looks at mediaeval and classical eras. However, what appears to be new is the concept of "science" as a separate entity or method through which we can systematically make sense of the world.

Defining religion

The concept of religion as a separate sphere of social life is itself a recent phenomenon that can also be traced to the renaissance era. Nongbri (2013) showed how the pre-modern Latin (and Greek, Hebrew and Arabic) terms now commonly translated as religion signified concepts that were subtly and sometimes not so subtly different from religion as conceived of these days. Defining religion and pre-modern concepts of what we would now call religion, he argues, is complicated by the fact that Western scholars have traditionally tended to define religion through the religious practices they are most familiar with. Most often, it thus tended to be Protestant Christianity that served as the standard through which religion was viewed and to which it was compared. Particular features of Protestantism were then taken to be defining features of religion in general, and religions in other countries and other eras are compared to see how closely they match up to these ideals in order to be judged as either religious or non-religious modes of belief.

The emergence and eventual establishment of Protestantism, itself a product of the renaissance and early modern era, was thus linked up with the intellectual development of a separation of how communities make sense of the world and how we should live within it. Science, secularism and the idea of separate religions modelled on Protestantism all sprang up as separate domains from a general but previously more interconnected human endeavour to understand the world, our place within it and how we should live. In this sense, it is not so much that either the establishment of science or the abandonment of religion leads to the secularisation of Western society, but that secularisation, science and religion are all co-evolving ideas that were the by-products of social changes and the developments of new "social imaginaries" of early modern Europe (Taylor, 2009).

Religion and science then can be seen sociologically as two sets of social practice, rather than particular collection of doctrines or facts. As Stenmark (2010) notes:

> In fact, the starting point for a more detailed account should be that science and religion are not merely sets of beliefs or theories plus certain methodologies, but are two social practices. That is to say, whatever else science and religion might be, they are complex activities performed by human beings in co-operation within a particular historical and cultural setting. We could roughly say that a practice is a complex and fairly coherent socially established co-operative human activity, through which its practitioners (for instance, religious believers or scientists) try to achieve certain goals by means of particular strategies.
>
> (Stenmark, 2010, p. 292)

Even then, though, the two terms clearly have meaning within contemporary discourse and tend to be seen and practised as separate realms of social practice, which can conflict, be in dialogue or even interrelate with each other but are nonetheless separate. Thus it makes sense to look at seeking to isolate some definitions of what these realms of social practice are that capture, in some sense at least, our intuitive understanding of them and what we are talking about when we refer to either science or religion.

Such attempts at definition are, however, fraught with difficulties and are highly contested. At root, one problem is that a definition would have to capture our intuitions of the concept while also being empirically relevant. Definitions that exclude practices or disciplines that we intuitively want to accept as religious or scientific will have difficulty being accepted. Thus any definition of religion that excludes, say, Protestantism would be suspect; similarly, so for any definition of science that excludes physics. The definitions would also need to exclude things we would intuitively class as not religious or scientific; for example a definition of science that includes astrology will find it difficult becoming accepted. What these "prototypical" sciences and religions are, however, is difficult to pin down, and the suspicion often arises that definitions are then merely worked out to conform with the definer's own preconceived ideas. Nongbri's (2013) argument that research on religion has traditionally often been unconsciously coloured by the researcher's own cultural upbringing (in an Anglophone academic environment, this is often some form of Protestantism) would be an example here: There is otherwise no good reason why, say, Siberian shamanism shouldn't be the prototypical religion after which Protestantism is judged. Durkheim (2008 [1912]) made a similar point regarding this difficulty of finding an acceptable definition of religion:

> If we are going to look for the most primitive and simple religion which we can observe, it is necessary to begin defining what is meant by a religion; for without this we would run the risk of giving the name to a system of ideas and practices which has nothing at all religious about it, or else of leaving to one side many religious facts, without perceiving their true nature.
>
> (p. 23)

A similar argument could be made of science: The social sciences are often judged as more or less scientific depending on how well they accord with the characteristics of a prototypical science like physics, however there is otherwise no good reason why we shouldn't judge physics' "scientificity" by how much it accords with the characteristics of, say, sociology.

It is thus almost impossible to have completely neutral definitions of either science or religion, as background cultural and social assumptions as to what needs to be included or excluded are already baked into them. Often, however, they are also coloured by theoretical (philosophical or sociological) positions taken up by the author: A definition of the main term is offered with the view of this lending support to wider theoretical points an author wants to make. Aldridge (2013) talks about Durkheim's own definition as a "Trojan Horse for his theory of religion" (p. 26), and I believe one could make that observation more widely covering most definitions of both religion and science (I would also invite the reader to read my own attempts at formulating a view on religion and science accordingly, as I don't believe I can do away with my own preconceptions and theoretical motivations).

Thus, though they are rarely put together like this, the problems of defining science and religion are remarkably similar, and in order to explore these similarities and arrive at a conceptualisation of social knowledge relevant to my project, I will very briefly review some influential attempts at definition to arrive at the current state of thought, which more or less holds that giving definitions is fairly impossible and probably not a particularly useful activity anyway. This is the case for both religion and science.

Popular definitions of religion might concentrate on the doctrinal content of the beliefs they hold. These might include concepts like God or gods, the supernatural or transcendence. One clear difficulty with these ideas is that none of them is itself clearly enough defined to be particularly useful. The supernatural, for example, which features prominently in new atheist discussions of what distinguishes science from religion, which stipulate that science looks only to natural rather than supernatural explanations, presupposes a category of "natural" explanations to which they are in opposition. But what do we really mean by natural and supernatural explanations? Once attempts are made at providing clearer definitions of these terms, we often end up having to count ideas traditionally associated with religion as not religious and vice versa, and we are back to the original problem noted above.

Social scientists have therefore sought to widen views of what religion is by going beyond definitions based on doctrinal content and including (or even focussing on) the social aspects of practices which make up a religious community. Like definitions based on content, these however also often face the problem that the central terms need further definition, and that when addressed we might end up with definitions that appear unintuitive. In his textbook introducing the sociology of religion, Aldridge (2013) distinguishes between inclusive and exclusive definitions, a distinction that mirrors this definitional dilemma. Inclusive definitions tend to cast their nets wide and as a consequence might need to include practices of beliefs that are not traditionally seen as religious, while exclusive definitions might end up excluding practices or beliefs that are traditionally seen as religious.

An early, inclusive definition of religion, and one of the most influential, came from Durkheim. Durkheim, as quoted above, recognised that defining religion needs to take into account a greater variety of religious experience than what we might be familiar with through our own religious backgrounds. Some traditional forms of Buddhism, for example, do not stipulate any gods, so any doctrinal definition that includes belief in a god will stumble here (see also Byrne, 1991). Thus Durkheim writes that religion is better seen as a social practice.

Defining religion through its rituals and social practices avoids the clear pitfalls of concentrating on the content of religious (and otherwise) beliefs. Talking about moral communities (churches), however, leaves out individual spiritual beliefs that we might also want to class as religious. Some authors distinguish between religion and spirituality on this or similar bases (see Hanegraaff, 1999). Social and practice views on religion can let us look for commonalities, even when there are otherwise no beliefs that are common to all religions. Focussing also on the emotional energies of religion, Durkheim famously wrote about the "collective effervescence" that crowds can feel when gathered together for a shared experience, an emotional energy that is "significant not only for the vitality of religious life, but for social life as a whole" (Mellor, 1998).

Ninian Smart, in his influential textbooks on world religions, worked with practice and social definitions of religion as his project clearly involved such a large variety of beliefs and religious viewpoints that any content-based definition would be inadequate:

> The search for an essence [of religion] ends up in vagueness – for instance the statement that a religion is some system of worship or other practice recognising a transcendent being or goal. Our problems break out again in trying to define the key term "transcendent".
>
> (Smart, 1989, p. 11)

Instead, he writes about seven dimensions, or aspects, of religion, not as a way of defining it but of accounting for its variability and helping social scientists describe them. These dimensions are (1) the practical and ritual, (2) the experiential and emotional, (3) the narrative or mythic, (4) the doctrinal or philosophical, (5) the ethical and legal, (6) the social and institutional, and (7) the material. These dimensions don't all need to be present to the same degree – or even at all – in a religion, however they help "give a balanced description" without "neglecting either ideas or practices" (p. 21).

This approach, as Smart comments at length himself, also gives the social scientist the tools to analyse movements or communities usually seen as secular. Humanism, Marxism and nationalism are all phenomena which can be seen through the lens of Smart's seven dimensions, along with football and pop-star fandom, even though they don't feature as part of the "world religions" that his textbook aims to introduce us to. Smart proceeds to provide an outline of how, for example, nationalism and the veneration of the nation state can be analysed in the same way that he then proposes to analysis movements more conventionally recognised as religious.

Nationalism, for example, includes shared symbolisms, ritual practices and social identities often bound together through charismatic and venerated leaders who anthropomorphically embody the wider concept of the nation (and/or its political system); there are shared myths and foundational narratives, and nationalism is often experienced emotionally (e.g. through the collective effervescence of a football match, a fascist campaign rally or going to war).

Thus, similarly to Durkheim's collective effervescence, the working characterisation of religion will include things not usually seen as religious, but then, now that we have the tools to analyse these large-scale human social, epistemic and ritual systems of belief and practice, the argument of where exactly to draw the line between the religious and the non-religious might not in fact be particularly important.

The few preceding examples are not intended to be a comprehensive review of this problem; variations of these attempts at defining religion are numerous (Aldridge, 2013; Furseth and Repstad, 2006), and while clearly an interesting literature to explore, the lack of consensus and conclusion is what interests me most here. The generally accepted working conclusion within the sociology of religion appears to be that maybe the lack of consensus on what religion is doesn't particularly matter in any case. If sociologists are interested in the practices, rituals and beliefs of a community, then they can study these well enough without having to conceptually force these into any artificial, enlightenment-era, socially constructed concept like religion, which due to its lack of coherence is explanatorily useless anyway (Byrne, 1991).

When forced, the discipline sometimes tends to view religion through Wittgenstein's (1953) concept of family resemblance (Byrne, 1991, discussed in Saler, 1999), where members of the conceptual family of religion resemble other members of the family in some respects, but not all, and it is perfectly possible to have two family members who don't share any features with each other. For this reason, Jong (2015) argues that "at best, [the terms of religion and non-religion] may retain their social functions – in names of departments, scholarly organizations, conferences, and journals, for example – but they have no legitimate scientific use" (p. 15). Jong also observed that

> social scientific research on religion (and related phenomena, including non-religion, atheism, and secularity) is invariably prefaced by sheepish attempts to define these terms, followed by apologies for the inevitable inadequacy of the proposed definitions.
>
> (p. 15)

Defining science

As with religion, we could possibly seek to define science by either its content (or the nature of its content) or through its practices. Looking at content is problematic because the general expectation is that science is always changing and thus the content of scientific knowledge is continuously updating. Nevertheless, for some purposes, such as scientific literacy surveys, or within science education, the idea that there is a corpus of scientific knowledge that one has to know in order to "know" about science has had a long tradition (e.g. Miller, 1998). While these are not really definitions of science as such, they clearly help delineate – for policy or educational purposes – who knows enough about science and who doesn't. These also can then be criticised because clear criteria between what counts as scientific knowledge and what doesn't can either be arbitrary or reflect the philosophical presuppositions of whoever sets the criteria.

In an education context, the setting of the boundary between science and non-science can lead to conflict, in some cases because of the enormous cultural prestige that science has, which will drive other disciplines to present their knowledge claims as scientific in order to be taught at all. In the science versus religion area in particular, this conflict can arise out of the secular philosophy that seeks to separate religion from state institutions, which for example drives school education in the US and thus leads to the long stand-off between the teaching of evolution and creationism.

In science communication studies, the use of science literacy surveys has come under criticism also, for leaving it up to whoever designs such surveys as to what constitutes proper scientific knowledge but also because of the more fundamental issue that merely knowing a lot of facts does not for many people constitute proper understanding of science if it is not accompanied with an understanding of the processes by which science arrives at reliable knowledge, and appreciation of the "scientific mindset" that allows science to be as successful as it has been (Gregory and Miller, 1998).

In any case, looking for the distinguishing features of science by appeal to the content of scientific knowledge is rarely attempted outside of educational or highly politicised contexts such as the evolution controversy. Instead, as alluded to above, appeal is made to the epistemological features of science, which allows us to arrive at knowledge that is as robust as possible. What exactly the "scientific method" is or should be has been a major preoccupation within the philosophy of science, and it is probably fair to say that this discipline is not in any danger of arriving at a consensus any time soon. The idea that there is a discernible "scientific method" which can be used as a demarcation criterion that will help us delimit science from non-science has now been more or less abandoned within the philosophy of science (see my brief overview in Riesch, 2014), with some philosophers now advocating pluralism in methodology instead (Chang, 2012).

While attempts at defining science through the elaboration of a scientific method following a particular logic have stalled, work in the sociology of science has focussed on the social practices of scientists both as a possible defining feature and as an epistemological argument for how certain (social) practices are conductive to arriving at reliable knowledge, somewhat mirroring the divide between belief and practice-based definition of religion outlined above. Robert Merton (1973) famously elaborated four social norms of science, which science follows – or should follow – in order to work: Universalism, communalism, disinterestedness and organised scepticism. Inspired by the philosophical work of Kuhn (1962) however, a group of British sociologists in the late 1960s and '70s argued that this conception of the sociology of science leaves any sociological explanations only able to account for scientists' behaviour, not the content of science itself (see Yearley, 2004). This leads to sociology explaining only wrong knowledge, or as they termed it, becoming a "sociology of error" (Bloor, 1973). Sociologists, however, should be agnostic as to the actual truth of the science and be able to analyse how knowledge is arrived at, regardless of whether it is true – this is something the sociologist is not in a position to pass judgement on. In addition, they argued that the social processes of arriving at knowledge would need to be the same regardless of whether that knowledge is true, a principle called the "symmetry principle" by Bloor (1991). This school of thought, variously called the Edinburgh school of sociology of science, the "strong programme" of the sociology of science or the "sociology of scientific knowledge", had a strong founding influence on the subsequent development of the sociology of science and the larger discipline of Science and Technology Studies.

With this development, the wider question of how science is defined has, similarly to philosophy of science, been abandoned – in this case not so much because it was realised that the problem is intractable, but more because the new sociological thought behind this did not really recognise this as a valid question: Sociologists should analyse "true" or proper science using the same explanatory frameworks as those beliefs that would otherwise be seen as not proper science, such as creationism, homeopathy or other non- or pseudo-sciences.

If there is a demarcation between science and non-science, sociologists have sought to find this within the self-presentation of the scientific community itself. An influential approach to this has been Thomas Gieryn's concept of boundary work performed by scientists (1999), by which he means the collective professional management of conceptual and professional boundaries that scientists engage in in order to exclude the others from the professional benefits they would derive from being part of science, in terms of prestige, funding and the opportunities to advance their viewpoints within the academy. Gieryn's concept of boundary work has been developed explicitly as an approach to the professional knowledge and community management engaged in by scientists. This is where it is applied to most often, however a similar theoretical approach can be made to the identity management of other more or less clearly delineated social groups (Lamont and Molnár, 2002; Riesch, 2010), and thus is not a distinguishing feature of science as such.

So again, similarly to the question in religion, neither philosophical nor sociological research on science has settled on any definition or demarcation between what is science and what isn't, either in terms of content, the logic of arriving at knowledge or the social practices of scientists as a group. The self-identification of scientists as scientists and the practices and strategies they use to define themselves have been fruitfully analysed, however these same practices can also be used by "non" scientists to police their own boundaries as scientific and have been, for example, in the case of scientific creationism. The best we can say about what science is will again probably be an idea similar to Wittgenstein's family resemblance.

Self-categorisation and the flow of ideas

The self-categorisation of people into clear social groups, be that as scientists or various religious communities, however, does show similarities which can and have been analysed through wider sociological and social psychological theories on social identity (Tajfel, 1981; Hogg and Abrams, 1988). The fact that the boundaries between science and religion (and other things that are neither) are socially contingent and constructed does not mean they aren't real to the people concerned, and this has real-world consequences both to how people live their identities and how they construct and arrive at the knowledge and knowledge practices that define them. Although national boundaries and nationalities are socially contingent constructs, they nevertheless constrain where I am allowed to live or travel.

In this sense, the two concepts of science and religion describe real social phenomena, and pointing to the lack of clarity in the definition of either, or the historical processes from which the two emerged out of a fairly unified quest for understanding about the world, does not mean the categories are useless (see also Nongbri's 2013 argument). Even when individuals or groups identify with both social groups (i.e. religious scientists), the two groups themselves tend to be socially separated as requiring these complex identities. To reprise the preceding example, even though my two nationalities are socially contingent constructs (I have dual British and German citizenship), identifying with either one, or both, or neither nevertheless has real-life consequences and is contingent on the cultural context that may view these identities as being in conflict (Germany does not generally allow its citizens to hold dual nationality outside of the European Union). Thus, in a cultural context (such as contemporary Anglophone countries) where religion and science are seen as in conflict, the conflicting identities will then be managed differently than in cultural contexts where the conflict is maybe not perceived as strongly.

This seems to be, then, the most sensible way of approaching the sociological study of both science and religion, not by treating them as essential categories which must fulfil one or more particular characteristics (either doctrinally, logically or in social practice) but as concepts that have grown organically out of our intellectual history and now conventionally form separate social, institutional and often cognitive realms.

The different social identities between science and religion (as well as between individual sciences and religions) shape the make-up and behaviour of the respective group membership as well as what the acceptable norms, values, concepts, modes of thinking and modes of evaluating experience are. This process pulls the social identities further apart, as outgroup membership becomes subject to stereotype projections that exaggerate perceived negative attributes, while ingroup membership enhances the perceived positive group norms and values held by an individual (more on social identity in relation to apocalyptic movements in Chapter 6).

Interrelation between the two, however, and the travelling of ideas, concepts or practices from one to the other is of course still happening and reflects the fact that science and religion share a common heritage. Usually these interconnections would be seen as inappropriate by those who hold to a "conflict" model between science and religion, or alternatively as a desirable common point on which dialogue or even integration can be started (in terms of Barbour's classification). However, these reprise the idea that there is an essential clear difference between the two which goes beyond the social identification of group members.

Instead it might be fruitful to see these as two (or more) sets of belief systems which in their own way attempt to make sense of the world, while keeping in mind the social contingencies which have shaped them into different group identities, and thus given different priorities to the different sets of epistemological practice that have emerged over time. The business for me as a sociologist is then to make sense of how we make sense of the world. Following the symmetry principle elaborated by the sociology of scientific knowledge (see above) then would also prohibit the analyst from presupposing which belief or system of beliefs is true, or more likely to be true, because the social (and philosophical/logical) causes of true belief should be the same sort of causes that cause false belief.

This is not to say that truth is not an important concept in how we do make sense of the world – I clearly have beliefs which I hold to be true, or at least approximately true, including beliefs about which epistemological practice is most likely to lead me to it. Similarly, while a completely leak-proof scientific, logical and empirical method of arriving at reliable knowledge has not yet been found, the approximation of this – or a good-enough fudge if you will – that makes up current scientific knowledge and practice more often than not determines my own beliefs about the world. These methods themselves are not completely nailed down yet by logic or philosophy, even if there is hope that this can be done in the future. However, the point is that the same kind of epistemological reasoning that leads to scientific knowledge claims can lead to knowledge claims in other domains too, and we can't generally use science-derived tools of evaluating these knowledge claims, because these evaluation tools are themselves knowledge claims in need of evaluation. Therefore they must also be analysed in a way that is agnostic as to the truth (of whether the knowledge claim evaluation is reliable), if our aim is to make sense of people's understanding. This, painfully, includes my sociological knowledge, through which I attempt to understand the social world of scientists.

With respect to my attempts to apply the symmetry principle from Science and Technology Studies in my analysis (and this reflexivity is in itself one of Bloor's other foundational principles for the sociology of scientific knowledge), it is interesting to note that the field of the sociology of religion has also elaborated similar principles that are intended to ensure that the sociologists' own beliefs and knowledge claims don't interfere with the sociological analysis of the beliefs they study. Peter Berger's so-called methodological atheism (Berger, 1967), for example, cautions the sociologist of religion to approach their study as if they were atheists themselves. The fact that the sociologies of religion and science have formulated these similar methodological principles seemingly independently of each other is probably an indication of the direction of travel towards the analysis of belief within the wider sociological discipline. However, to me this also indicates that the sociological analysis of these two areas of social life have unreasonably lost sight of each other, and each had to invent their own version of the wheel.

With all this in mind, I would argue that some of the ideas, concepts and social practices that social scientists have noted as shared between science and religion, either through common origin or travel between them, should not be caricatured as "science being a form of religion" or (which I hear less often) that "religion is a science". Instead, I mean to show that ways of understanding or making sense of the world, whether in a doctrinal, a theoretical or a social, practical way can be remarkably similar and analysed as such. To develop this, in later chapters I will introduce discussions about scientists' rhetoric acting analogously to prophetic discourse (Walsh, 2013), that the idea that "Nature" is a non-theistic continuation of the concept of God and that "risk" has come to fill the explanatory gap left when secularisation abandoned the concept of fate.

Ian Barbour, introduced above as the towering figure in the scholarly debates on how religion relates to science, has engaged in a similar attempt to show parallels between scientific methods and the logic through which religions arrive at knowledge claims through the claim that religious communities can be seen as analogous to Kuhnian paradigms. This has been criticised (Berg, 2002) and may at the end have been a failure, certainly if we take Barbour's aim to be showing how the reliability of scientific knowledge claims can be transferred to the religious realm.

However, seeing religious movements and/or communities as analogous knowledge communities to scientific paradigms may make sociological, if not theological, sense if re-interpreted through social identity theory as socially cohesive epistemological ingroups whose social norms, values and knowledge claims are shaped through their collective self-categorisation as well as their categorisations of the outgroup(s). With the proviso that Kuhn's paradigm concept as such has been refined in philosophical literature through concepts such as Lakatos' (1978) research programmes and, in sociological literature, the boundary work of self-defining scientific communities, we can look at the social identity work through which both religions and scientific communities define valid knowledge, valid epistemologies for arriving at knowledge, and finally their definitions of ingroup members who are allowed to pronounce valid knowledge claims as analogous (i.e. working through the same underlying social psychological dynamics) without needing to make a methodological commitment to the truth or falsity of the actual contents of their knowledge claims. Analysing the science of, say, climate change as arising through similar social psychological causes as UFO millennialism does not mean I cannot commit to the belief that science has hit on an epistemology that is broadly right while UFO apocalypse theory has its epistemology wrong. It is in this spirit that I hope the analysis in the following chapters will be taken.

This chapter was intended to serve the purpose of introducing the main relevant concepts from the philosophy and sociology of the science/religion relation, as well as the sociologies of religion and science more broadly. An attempt such as that of this book to analyse both religious and scientific conceptualisations of the catastrophic near future faces a decision on whether to use sociology of religion theory as the main source of analysis and fit it around science, or vice versa. Given my own background, the temptation to make this a Science and Technology Studies analysis of religion is strong, and no doubt this bias will shine through in several places. The alternative is that we might need to look at connections between these two branches of sociology as a way of building a theoretical framework that can deal with both social realms. As I have tried to show in this chapter, the connections between religious and scientific communities, identities, thoughts and practices are close enough for a more generalised sociology (and social psychology) of knowledge to make sense of how we make sense of the world, and I will try to elaborate such an approach in the following chapters.

Chapter 3

The meaning of death and the making of time

Introduction

Why do we keep talking about the end of the world? Given the long history of apocalyptic predictions of an imminent end, none of which has so far been borne out, one overarching research question of sociologists and historians studying millennialism is to figure out how the fascination about the apocalypse comes about. The end of the world as a general proposition can take a lot of different forms, and certain strands and themes that guide us into how we think about the apocalypse have become part of (at least Western) social and cultural heritage, and form part of our social knowledge.

A society's social knowledge is not created from scratch or in a vacuum; it evolves out of previous knowledge and understandings of the world and interacts laterally with other societies' social knowledges. The process of change in social knowledge is mostly gradual and imperceptible, however times of crisis and revolution can punctuate the evolution of our social knowledge more dramatically, comparable maybe to the conceptual revolutions of philosophers like Kuhn (1962) diagnosed within the history of scientific thought. As I argued in the previous chapter, a clear distinction between a society's scientific knowledge and its wider cultural understanding of the world that is its social knowledge is not quite as clear-cut in any case. However drastic these changes though, they probably only appear as drastic with hindsight and may not be so obvious to the people living through them.

In order to arrive at a wider conceptualisation of the contemporary apocalyptic in later parts of the book, this chapter will first outline a brief history of apocalypse and apocalyptic thinking, ending with a brief overview of some of the social theories that have been advanced to make sense of them.

The making of time: a very brief history of the end

As with all social knowledge, direct threads and lines of influence can be drawn from earlier beliefs. Millennialism and apocalyptic worldviews can be traced back to other Middle Eastern religions and cultures, just as I will argue that the

Christian ideas they gave rise to have in turn influenced contemporary understandings of the end of the world. Ancient Judaism developed at the confluence and under influence of other cultural traditions in the wider area, and influenced them in turn (Gnuse, 2011). Most prominently, these were the Egyptian and Mesopotamian civilisations, and the Zoroastrian religion of Persia. Later on, with the Macedonian conquest of the Middle East, Hellenistic civilisation and later still Roman civilisation also influenced it, though possibly more as an outsider conqueror the elimination of which apocalyptic hopes were often set on.

As possibly foreshadowing later millennial ideas that imagine an apocalyptic clash between good and evil followed by redemption and a reign of peace and righteousness, Gnuse (2011) writes that some ancient Egyptian prophetic texts envisage a societal breakdown followed by visionary pharaohs who will restore the country to a golden age. Later on, when Egypt faced its own struggle with Hellenistic and Roman rule, some Egyptian texts describe almost millenarian ideas on new saviour pharaohs specifically saving the country from foreign rule. Likewise, Gnuse describes a series of early Mesopotamian texts and prophecies that speak of times of disorder and destruction followed by the emergence of a strong ruler who will defeat his enemies and bring peace to the land, thus ushering in a golden age. These are motives that he argues have become part of the Jewish and thus later Christian "language of the Millennium" (p. 239).

More difficult to trace, because of the absence of contemporary written records, is the influence of Zoroastrianism on Jewish millennial thought. Although that religion has been around for a long time and would have clearly been able to influence the development of Judaism, this religion was transmitted orally and religious texts were not written down until the after the Muslim conquest of Persia. Parallels we might find between Zoroastrian and Jewish texts regarding the apocalypse and the millennium therefore might as much indicate the influence going the other way round. There are some very strong similarities that have traditionally convinced scholars that Zoroastrianism did have a great influence on the development of Judaism. Kreyenbroek (2002) for example cites the similarities between Zoroastrian ages (with world history being divided into three to four ages of 3,000 years each, and with the current generation living towards the end of the last one) and similar visions in Daniel 2:31–45 (KJV) as having been used as evidence, even if modern scholars are seemingly less convinced. There are also other similarities, for example the belief in a saviour figure.

History is then understood as cyclical in a sense but also hurtling towards a clear and defined end, still far in the future but nevertheless a decisive and final victory of good over evil. This judgement day was a relatively new view of the world within contemporary religions. As Gnuse argues, while Egyptian and Mesopotamian prophecies envisaged final events and final days, the righteous triumphant and foreign oppressors vanquished, these were very much temporal events which would happen in this world and not as part of a greater plan to visit judgement after death and after which time itself will stop. Thus the periodisation of history into separate ages, the "imagery of woes at the end of time", bodily resurrection, the concept of personified angels and demons. and finally the personification of evil "as a supernatural force" (p. 247) are all innovations of the developing Middle Eastern eschatology of early Judaism and Zoroastrianism.

Just as crises and foreign invasions seemed to give rise in Egyptian, Mesopotamian and Persian popular imaginations to hopes about future redemptions and native restorations, so they did in ancient Hebrew civilisation. Millennialism as it developed in that society would in due course become one of the most influential ways of making sense of the end, as developments of these ideas would of course then spread through the expansion of Christianity and later Islam throughout Europe, the Middle East and beyond. Tabor (2011) argues that Jewish millennialism developed as a response to the conceptual problem of evil (cf. O'Leary, 1994), with a particular emphasis on the turmoil and successive invasions (Mesopotamian, Greek and Roman conquerors, the Babylonian exile) the region went through. How could a good and powerful God allow such suffering? It was especially the religious response to the Maccabean revolt against the Hellenic rule of Antiochius IV and the aftermath, "as formulated by the author of the book of Daniel", that Tabor argues "shaped the parameters of most forms of apocalyptic Judaism, including early Christianity, in the centuries that followed" (Tabor, 2011, p. 255). Thus the apocalyptic prophecy of Daniel appears to have been written as a supposedly more ancient prophetic text detailing contemporary events.

As Tabor outlines, the book of Daniel has become "foundational" to Jewish millennialism; it contains four essential elements that later apocalyptic thought carried over, and this is part of any subsequent Jewish, Christian and Islamic account of the end. These are the expectation and imminence of the end times, that a series of signs will show us that the end times are near, a clear chronological scheme of what events will happen in the lead-up to the end and finally the final judgement of both still living and already dead (Tabor, 2011, p. 256).

All of these clear prophecies did of course not come to pass, as Tabor points out, the Maccabees were successful in their revolt but the country would shortly after fall to the next round of invaders (the Romans). However, despite prophecy failing to come true, "the book of Daniel was grasped all the tighter to the collective breasts of those Jews and early Christians who, unwilling to accommodate to the cultural hegemony of Roman rule, continued to hope for God's intervention"; the "millenarian rescue" was always "just around the corner" (Tabor, 2011, p. 257).

There was a great variety of apocalyptic and millennial expectations in ancient Jewish and early Christian thought. As Tabor points out:

> There are texts that focus on messiah figures, while others have no messiah at all; tests that specify a period of time and those that have no such designations; and tests that describe the reign of God in earthly terms as well as those that envision and transcendent heavenly state.
>
> (2011, p. 253)

Early Christianity can be seen, as introduced above, as one of many varieties of contemporary Judaic apocalypticism. The future messiah who was expected to lead the humanity into the last days (as outlined in Daniel 9 and several other non-canonical documents such as the books of Enoch; Collins, 2002) was seen in several charismatic figures and leaders. At the beginning, apocalyptic sects that clustered around a messianic leader start with a very clear expectation that they are now living in the end times and that judgement day is near. So far, of course, all these expectations have been frustrated, and a movement that wants to survive the disappointment of the world failing to come to a conclusion must put some work into reinterpreting these expectations.

Although Jesus' was one of many millennial nationalist movements, Attridge (2002) argues that Jesus used contemporary Jewish images of the reign of God, derived possibly from Daniel and the Psalms of Solomon. This was a royal power that stood in direct opposition to the imperial (Roman) power dominating Israel at the time, and he was therefore seen as a direct threat to the authorities, more so than maybe some of the other contemporaneous prophets.

As Festinger, Riecken and Schachter (1956) have argued (see Chapter 5), disappointed prophecy rarely leads to abandoning of the belief; instead the belief system can easily be tweaked through some additions that reconcile the belief with how events have turned out. In the case of the apocalyptic sect surrounding the famous Dead Sea Scrolls, which expected the imminent arrival of the messiah, the last scrolls from around 40 years after the earlier prophesised messiah failed to turn up, prolonging the approach of the end into some time in the future, where "there shall be yet another vision concerning the appointed time" (quoted in Tabor, 2011, p. 261).

In the case of the sect surrounding Jesus, ostensibly his death through the authorities would have put an end to expectations that he would lead humanity into the final days; consequently the belief system was upheld through the additional belief that his death was far from final and that he would instead rise from the dead and carry out his apocalyptic function. This second coming had initially been expected to occur fairly quickly. In a pattern that has been observed very regularly in apocalyptic movements whose predicted end-of-days date has passed without incident or otherwise been frustrated through the death of the person supposed to play a starring role, the date has been shifted into the not so distant future. If a precise revised date has been given, the movement is bound to suffer another disappointment. If not, as was the case with the group of followers that Jesus has gathered, the years can go by and by before a second disappointment happens, but nevertheless the belief system will need further adjustments as it becomes more and more settled into a new normalcy that, similar to the Dead Sea Scrolls, postpones the final event into a more unspecified future date (or even into some unspecified, allegorical time).

For the first few generations after Jesus' death, the emerging Christian community still expected a fairly imminent second coming and associated end of days. The visions that John of Patmos wrote down in Revelation (around 80–100 CE) were still very much expecting that the end would come any day now; the signs and symbolism he wrote that herald the end are easily interpretable in terms of the political situation of his day, with the hegemonic Roman Empire taking centre stage in how the last days would develop.

As Eugen Weber (1999) puts it, St John's imagery in Revelation, with the four horsemen, the seven seals, trumpets and vials, the Antichrist beast seducing and subjugating nations and waging war, the earthquakes, locusts, pests and falling stars have infused the popular culture of succeeding generations, providing them with "metaphors, allegories, and figures of speech, with warnings, guideposts, and inspiration" (p. 30). The precise sequence of how events are supposed to happen and the precise interpretations of the meaning behind the various imageries is a matter of much theological and historical debate that is impossible to get into here. However, the main narrative incorporating the apocalyptic influence of Daniel that Tabor noted as features of most post-Daniel apocalyptic writings – signs and portents of the end, a clear chronology, judgement and imminence of the end – are there, as well as the second coming of Christ as a warrior, the defeat of the Antichrist and Satan that binds him for a thousand years (the "millennium"), before he is loosened again to be defeated in a final battle that will finally bring about judgement, God's kingdom and everlasting peace, where "God shall wipe away all tears from their eyes; and there shall be no more death, neither sorrow, nor crying, neither shall there be any more pain: for the former things are passed away" (Revelation 21:4, KJV). The chronological order of the second coming and the thousand years during which Satan is bound is a matter of large disagreement between different theological schools and denominations. Traditionally a distinction is made between premillennial interpretations that expect the second coming to occur before the start of the millennium and post-millennial interpretations that expect the second coming to happen at its end (and which are thus often claiming that the millennium has already started). This distinction tends to align with the theological interpretation's general outlook: As Wessinger (2011a) notes, premillennialists tend to be pessimistic about human nature, whereas postmillennialism "involves belief in progress into the collective salvation" (p. 721).

However, as the new religion established itself and evolved from a cult to a minority to finally a state religion, expectations around the imminence of the second coming became a less and less important feature. The nature of apocalyptic longing (linked to the apocalyptic's origins as the oppressed nations' nationalistic dreams of rebuilding) is bound up with the clear minority status of the movement, where the apocalypse will finally vindicate the faith of the minority against the evils of the world around it. When the apocalyptic religion moves from being a minority religion, or one linked to a specific national identity, into a large, universal and temporally powerful force, then the longing for an apocalypse that sweeps aside the wicked world becomes problematic because now the powers that be are part of the religion. One way out of this is a post-millennial interpretation that believes the millennium is already under way, and that the establishment of an all-encompassing church means that Satan is at least temporarily banished.

But another way, sometimes called "amillennialism" (Wessinger, 2011a), formulated in the toweringly influential 4th- and 5th-century work of Augustine (1972), married the various and often contradicting scriptures into an interpretation that allowed theology to move forward into what Landes (2011) called "normal" rather than "apocalyptic" time. Under Augustine's influence, church doctrine thus became formulated in ways that accorded the end of days – and the biblical passages thereof – a more allegorical meaning, or as Weber (1999) puts it:

> Augustine delivered what looked like a knockout blow to literal readings of the Apocalypse: God's timetable was inscrutable, the city of God lay in heaven, the city of man lay on earth and never the twain would meet. . . . the Apocalypse is an allegory that shows the way to salvation, but it was silly to get excited about the millennium.
>
> (1972, p. 34)

The move from apocalyptic sect or cult to a decidedly non-apocalyptic majority religion then makes sense, especially if, as was the case with Catholicism, the religion takes centre stage in developing the moral and bureaucratic infrastructure of a whole continent, since in Augustine's thought the established Christian church is "equivalent to the millennial kingdom" (Wessinger, 2011a, p. 717). An apocalyptic movement hopes for the end of days because the world is fundamentally wicked, countries and the authorities that run them are corrupt and evil – it is this malaise with the world that the apocalypse is supposed to wash away. This argumentation is eminently plausible when the cult itself is opposed to the power structures of the day but becomes more problematic when the cult metamorphoses into a majority religion: It cannot blame the establishment anymore because it has become part of it. This move away from literal apocalypticism to allegorical interpretations as religious movements establish themselves as major religions in their own right appears to be a trend; similar moves can be seen in the history of Islam and the Protestant denominations.

Since even with the establishment of the church as an institutional majority religion the world in general continued to be wicked and evil, literal apocalyptic ideas and movements never really went away but instead now started seeing the established church as part of the evil world the apocalypse will sweep away. Literal apocalyptic interpretations of scripture were resurrected periodically and especially in times of turmoil, influentially so for example through the 12th-century writing of Joachim of Fiore (Reeves, 1980). The history of Christianity is pocketed with the periodic appearance of prophets, self-proclaimed messiahs and their associated apocalyptic cultic following, who exploited similar popular sentiments and hopes on an imminent end to the world followed by the final judgement that sustained downtrodden populations (Weber, 1999; Cohn, 1957). Again, similar trends and apocalyptic splinter groups can be seen in the history of Islam, once that religion had established itself (Bashir, 2002; Filiu, 2011). Just as the established Jewish belief system and community spawned many varieties of messianic and apocalyptic sects that led to Christianity, this trend continues down the generations of new religions. After the Reformation Protestant apocalyptic sects had established themselves, they lost the millennial fervour, a process which in turn motivated the formation of new apocalyptic Protestant denominations such as Pentecostalism.

With the discovery and European settling of the New World, conceptions of starting a new life and a new community among the empty lands abounded (though of course they were decidedly not empty). Ideas of the Americas as the millennial paradise alluded to in scripture began with Columbus himself (Watts, 1985 cited by Wojcik, 1997, p. 21) and continued with the Puritans and other escapees from the European religious turmoil induced by the millenarian struggle between Reformation and Counter-Reformation as well as the struggles for political supremacy among the reformation millennial movements. America, for the Puritans, represented a new beginning and a stage at which the millennial kingdom and Christ's second coming – events in which they of course were to play a leading role, chosen by God – could come to be. In the 17th and 18th centuries, many charismatic preachers and prophets of Puritan and Calvinist heritage maintained "that the world is unredeemable, and that human salvation is predetermined by God" (Wojcik, 1997, p. 22), and these teachings were supplemented with apocalyptic imagery (p. 24).

The period in the early to mid-18th century in American religious thought is often called "the first Great Awakening" (the first of a series of Great Awakenings characterised by heightened religious sentiments; a second occurred in the early 19th century). This movement established a mostly postmillennial (see Chapter 1) tradition that viewed America as "having the prophetic destiny as a chosen nation that would redeem the world and usher in the millennium" (Wojcik, 1997, p. 24). This millennium was, as is usual in apocalyptic imaginations, eagerly anticipated as close and imminent, and periods of worldly conflict (such as the revolution some years later) were eagerly interpreted by some as signs that it's all about to start.

The first two Great Awakenings were, despite what the name might suggest, not uniform movements but rather a collection of different groups, sects and interpretative communities that, while they shared many characteristics, also competed fiercely against each other. The late 18th to 19th centuries saw a proliferation of religious groups with millennialist views seeking to build their own utopias, often through shutting themselves and their communities off through communal living, a development echoing the communal millennialism observed by Cohn (1957) in late mediaeval Europe of, for example, the Taborites. Prominent among these were communities like the Christadelphians, the Oneida Community and the Shakers.

Other preachers were less interested in establishing select millennial communities and sought instead to reach a wider audience in preparation for the (imminent) coming of the end. The trajectory of William Miller's prophetic career of predicting the end to occur in 1844 and seeing that date pass without incident (dubbed "the Great Disappointment") is often cited as a classic case study in millennial preaching, proselytising and, ultimately, failure (and as a case study in failure, it will be revisited again in Chapter 5).

The apocalyptic continued to play a role in the development of US evangelical churches after the failure of Miller's prophecy, first through the survival of Miller's movement as the Seventh Day Adventist churches and then later through the development of new radical evangelical movements such as Pentecostalism or the Jehovah's Witnesses. While there is great variety within these new evangelical churches, a particular premillennial form of theology developed that over the 20th century developed into a more or less coherent, literalist interpretation of the near apocalyptic future. Taking on some of the ideas of British preacher John Darby (1800–1882) that the world history can be divided into discrete ages or "dispensations", with us living near the end of the last one, this tradition developed into a set of related theologies often denoted as dispensationalism or dispensationalist premillennialism (Boyer, 1992; Wojcik, 1996). While these churches were more marginal during the first half of the 20th century, they managed to assert influence more widely towards the end of the century and are currently politically very influential, although still numerically a minority within American Protestant denominations. Their apocalyptic influences and theologies can probably easily be overstated, and there are many other theological distinctions; as Boyer notes, the prevalence of these churches and the numbers of their congregations is not necessarily an indicator of how many people really believe that the end of the world is imminent or any of the other often very specific apocalyptic prophecies. Nevertheless, a fairly uniform set of beliefs about the imminent end has become formulated and spread widely through a series of prominent prophets and preachers utilising traditional magazine and book publishing to reach their audience but also finding new means of dissemination through radio and tv shows, fictionalised treatments of their apocalyptic prophecies (the *Left Behind* series of novels by Tim LaHaye and Jerry B. Jenkins, published between 1995 and 2007) and more recently the internet.

These prophecies relate to an often very specific reading of Revelation that connects the narratives the book outlines with contemporary political events. One of their most influential publications, Hal Lindsey's (1970) airport bestseller *The Late Great Planet Earth*, outlines some of the main strands. He notes for example that the temple in Jerusalem will need to be rebuilt before the events of Revelation can start, which often gets widened to a restoration of the wider Israel within its biblical borders (which explains the enthusiasm of evangelicals for Jewish control of Jerusalem, the Temple Mount and the Palestinian territories). Revelation notes that the Antichrist will lead a coalition of ten nations and comes from, or is connected with, Rome. This is generally thought to be historically referring to the Roman Empire at the time of St John, who after all expected the end times to come very soon, but in a premillennial interpretation need to be seen to refer to contemporary or near-future events. Thus political structure of the European Union (initiated after all through the Treaty of Rome in 1957) is often equated with the union of nations the Antichrist will come to lead (although Lindsey, but not all of the other premillennial prophets, declined to identify any particular political actor as the Antichrist; see also Knowles, 2018). The nations of Gog and Magog that will attack from the north are identified as the Soviet Union, noting the similarity of names Magog and Moscow and the fact that Moscow lies almost exactly due north from Jerusalem. These and many more parallels, some more imaginative than others, have been found between the events related in scripture and the contemporary political situation (see Boyer, 1992; Dittmer and Sturm, 2016) to argue that the signs point towards the imminent period of tribulation (from which the faithful will be spared through the "rapture") and the subsequent events that lead to the day of judgement. Although there are a great variety of different dispensationalist preachers and prophets with their own variations and differences of emphasis; a lot of cross-fertilisation and communication has created a more or less coherent apocalyptic picture. Interestingly, this cross-fertilisation has also influenced contemporary Islamic apocalyptic movements, with groups like the Taliban, al-Qaida and more recently the Islamic State taking up many of the ideas from US evangelical end-times preachers and incorporating them into their own millennial theologies, sometimes even quoting them directly, creating what Filiu (2011) described as a "fascinating sort of mirror effect" (p. xii).

The millennial movements, sects, cults, denominations and religions outlined above all derive from the Zoroastrian/Abrahamic Middle Eastern religious tradition, and this tradition is usually the main scope of studies in millennialism or apocalypticism. Of course, other religious traditions have their own eschatologies and ideas of what lies in store for us. Distinctions can be made between different phenomenologies of time (Eliade, 1954), and particularly between linear and cyclical conceptions, however even in cyclical conceptions of time the transition between cycles can also often noted as catastrophic, and apocalyptic in a wider sense. The extent to which these would be "properly" apocalyptic or millennial or a different phenomenon depends largely on how tightly we define these two terms, and just as with religion in general, the temptation exists to class other eschatologies in accordance to how much they conform with or deviate from our own (Protestant) notion of apocalypse. Thus, it is not entirely clear whether the apocalyptic is a clearly Abrahamic phenomenon or a wider phenomenon. This is all the more complicated by colonialism and the cultural globalisation through which Christian and Islamic theological worldviews have spread throughout most of the globe. Some of the most well-discussed case studies of non-Western or Islamic apocalyptic movements, such as the Xhosa cattle massacre (e.g. Landes, 2011 ch. 4; Steyn, 2000), the Taiping Rebellion (Lowe, 2000) the Sioux Ghost Dancers (Pesantubbee, 2000) or the Papuan cargo cults (Trompf, 2011; Landes, 2011, ch. 5) have all to some extent developed under the influence of Western religious tradition. In any case, Wessinger's (2011b) collection included several chapters on Far Eastern, South Asian and African millennial movements.

Another important set of apocalyptic movements often discussed within the study of millennialism but not usually associated with religion are the large rebellions, revolutions, military coups and philosophies such as Marxism, Nazism, and the French and the various communist revolutions. For example Barkun (1974), in his influential book on millennial movements (see more below), explicitly included a discussion on rebellions and revolutions. A revolution has many similarities with millennial movements in that it agitates towards a complete resetting of power structures, with an underlying philosophy that promises a new beginning, a utopian millennial kingdom and a clear expectation that judgement will be dealt to the evil oppressors the revolution seeks to overthrow. While all revolutions and rebellions are different in their own ways, Barkun notes that the "ritualised, consensual rebellion" and "millenarian revolution" lie on different ends of a spectrum, and that "the distinction between revolution and millenarianism is thus largely artificial" (1974, pp. 22–23). We may hypothesise that the general Christian cultural background social knowledge (about earthly millennial kingdoms brought about by violent cataclysms, with clear ideas about good, evil and their coming judgement) may have influenced even the nominally secular, humanist and atheist revolutionary movements in the formulation of their philosophies. The parallels between the French humanist republic or the various socialist and communist utopias with the millennial kingdom of scripture is often striking, as is the violent cataclysmic revolution that is needed in order to bring these about. Not often made explicit, these parallels are however occasionally drawn directly, for example Hitler's vision of a "thousand-year Reich" was a direct evocation of biblical language (the term "third Reich" appears to have been an allusion to Joachim of Fiore's third and final millennial state; Redles, 2011).

One important thing to note here is that, analogous to the continuously failed expectations of an apocalyptic end to the world, these and the more religiously orientated millennial movements discussed earlier have likewise never managed to build their millennial kingdom: The experiments in perfect societies have either turned sour quickly or never really worked as a utopia for everybody, as one person's utopia can easily become another one's dystopia (see Chapter 7). And certainly, they rarely last anywhere close to a thousand years (unless, I suppose, we turn back to Augustinian amillennialism). The cataclysm often inherent in both bringing about the New Jerusalem and the almost certain disappointment inherent in running is one of the reasons millennial movements have become perceived almost proverbially violent. The connections between millennialism and violence (explored for example in Wessinger, 2000) has several roots. One is that millennial movements are, as Barkun (1974) notes, on a spectrum with rebellion and revolution. The fact that they also tend to posit a clear delineation between good and evil orientates the movement very clearly and decisively against the outgroups who are now not only not just "one of us" but actually evil and to be resisted. Millennial movements always tended to be movements by outsiders and the marginalised (sometimes rooted in nationalist revival projects), as outlined above: They are a reaction to an unjust and wicked world usually by the oppressed (and/or their sympathisers).

Even when millennial movements succeed in taking over control in their society, their outsider self-categorisation leads to never really accepting the fact that they are now in control and seeing nefariously controlling members of the status quo behind every curtain: The Jewish world-order conspiracy that drove the Nazi victim complex; the behind-the-scenes aristocratic or bourgeois counter-revolutionaries that motivated the guillotine in the France and the Stalinist purges in Russia; or to take a more "traditional" religious millennial movement, the reign of terror instigated in John of Leiden's Anabaptist New Jerusalem in 16th-century Münster (Cohn, 1957). In all these cases, the movement perceives itself as not in control even where it is, and any imperfections in the new world they are trying to construct must the fault of the enemy. Seeing enemies and agents of the old order behind every curtain is one of the reasons the millennial kingdoms descend into dystopias so quickly.

Disconcertingly, seeing that millennial political coups, rebellion and revolutions lie on a spectrum and need not always have *explicitly* formulated millennial philosophies, we can also see parallels here with contemporary political projects such as Trumpism and Brexit, where the movement has somewhat unexpectedly found itself in power, is trying to implement a new utopian society that previously only existed in their own incoherent philosophy and – upon seeing this project fail to become the New Jerusalem they promised themselves – blame and persecute the representatives of the old order for thwarting and sabotaging the project.

With all the similarities noted here between the various ancient Middle East, Islamic pre-reformation, Protestant and humanist varieties of millennialism, there also exists of course a lot of variation between them. There are differences in the ideas on what the apocalypse will consist of and how severe it will be (e.g. from a literal end of the world to the "mere" revolutionary cataclysm; see Chapter 4), or when exactly the apocalypse will occur (although of course, this is usually somewhere in the very near future). Another difference would be the nature of the post-apocalyptic, which in turn (among other things such as conceptions of the nature of evil) influences whether the movement is actively looking forward to the apocalypse with dread or eager anticipation.

Thinking about the role of the apocalypse, the nature of humanity and its salvation that characterises the doctrinal differences between pre- and post-millennialism, Wessinger has influentially categorised millennial movements as either catastrophic or progressive: A catastrophic movement is pessimistic about human nature, believes the world needs complete destruction in order for the millennium to arrive and believes in a simple dualistic contrast between good and evil (equivalent in Christianity to premillennialism Wessinger, 2011a, p. 718). A progressive movement, by contrast, is optimistic about human nature and believes "that the imminent transition to the collective salvation will occur through improvement in society" (p. 721), and they may not necessarily believe a catastrophe is necessary to bring about the millennium (equivalent to postmillennialism; Wessinger, 2000, p. 8). An avertive millennial movement is added by Wojcik (2011), which believes the apocalypse can be prevented from happening through spiritual practices and faith, for example.

The meaning of death: theorising the end

The next sections will review some of the theoretical approaches to the apocalypse that have been written in the past. These tend to reflect the disciplinarily influenced theoretical preoccupations of the theorist and as such often address somewhat different aspects of the apocalyptic and apocalyptic movements. The religious scholar might be interested in what drives the religious aspects of apocalypticism, the political scholar might be interested in the social impact of apocalyptic movements, and the rhetorician might be interested in the rhetorical strategies and narratives that surround how people talk about the apocalypse. Thus this review will intend to provide an overall view of what we know about the topic and will argue that they are complementary and at times in opposition to each other.

Relative deprivation theory and its critics

As a concept from the wider sociology of religion, this idea concerns itself with the reasons that attract people to religion in general. According to relative deprivation theory, the answer is the relative levels of deprivation in a given society. People of marginal status relative to the norm within that society will be more attracted to religion, because religion (most often) teaches social egalitarianism, eventual just rewards for people living the good life in the hereafter, and likewise punishment for the not so good (in the Abrahamic religions at least).

This approach is sometimes traced back to the views of Karl Marx, who famously described religion as the "opium of the people" (*Contribution to the Critique of Hegel's Philosophy of the Right*, in Marx, 1978), which thus appeals to those lower down the hierarchy and to make them accept their lot in society through the promise of justice after death. Glock and Stark (1965, described in Furseth and Repstad, 2006, pp. 111–114) distinguish between five types of deprivation: Economic, social, organismic, ethical and psychic.

Relative deprivation theory covers religion in general, but how suited is it to explain in particular apocalyptic and millennial movements? Norman Cohn, in his classic history of mediaeval and renaissance millennialism in Western Europe, contends that relative deprivation is one of the main attractions of the millennial movements he looked at – the other, related one being that the movement can provide believers from the margins of society with a unique, special and important place in the world:

> For what the propheta offered to his followers was not simply a chance to improve their lot and to escape from pressing anxieties – it was also, and above all, the prospect of carrying out a divinely ordained mission of stupendous, unique importance. This phantasy quickly came to enthral them in their turn.
>
> (Cohn, 1957, p. 318)

The appeal of the millennial movement to the marginalised is also enhanced through the usually humble origins of the prophets. The prophet, as Max Weber defined him (and it is usually him), is "a purely individual bearer of charisma, who by virtue of his mission proclaims a religious doctrine or divine command-ment" (1963 [1922], p. 46). One aspect that distinguishes them from clergy is that they "propagate their ideas for their own sake and not for fees" (p. 48); the prophet usually comes from the laity (Weber, 1963 [1922], p. 79; Cohn, 1957, p. 41), thus signalling they are not part of the religious establishment.

While superficially at least relative deprivation theory makes sense, it has been criticised because the empirical evidence is not quite conclusive, and if there is an effect it would at best be marginal. Stark (1964) tries to explain the decline in religiosity in the working classes in post-war Britain by noting that "radical politics" may have become "a more attractive outlet than religion for their status dissatisfaction" (p. 698), which may however also be an argument for viewing radical politics as fulfilling religious social functions: Saving relative deprivation may require us to rethink religion, so it may not be the solution for anyone with a more restrictive idea of what religion is.

Further, as Furseth and Repstad argue, various theoretical criticisms can be made, including that the theory only focusses on factors that make religion attrac-tive but neglects the effect of proselytising by its members, or that to make a clear commitment to a religion one actually often needs financial resources (Bibby and Brinkerhoff, 1974).

Disaster theory

Arguing that relative deprivation doesn't manage to explain the full spectrum of millennial movements, Barkun (1974) contends that millennialism and apoca-lyptic movements are mainly caused (or at least facilitated) by disasters, both natural and man-made. Barkun departs from the question of "why . . . millenar-ian movements [have] occurred where and when they have" (p. 34). While he argues that relative deprivation theory makes sense to an extent, it also misses important things. For example, he argues that it involves an essentially subjective judgement (where you stand with respect to general society), and people don't rationally look at their exact situation and then decide whether to join a millennial movement. So whatever is going on, relative deprivation and similar explana-tions can only provide a rough outline of the appeal of millennialism (and reli-gion generally); they can't account for individual actions. But more importantly, he also feels it fails to provide a "full explanation" of millenarianism, because in some situations relative deprivation does indeed seem to produce it while not in others. Thus he feels that there needs to be something more in the explanation; relative deprivation is merely a necessary but not a sufficient explanation (p. 37). Instead, what he observes to be a common theme in millenarian and apocalyptic movements, with reference to the substantial historical work of Cohn (1957) and others, is that "they always occur in times of upheaval, in the wake of cultural contact, economic dislocation, revolution, war and natural catastrophe" (p. 45). Vice versa, he argues, millenarian movements tend not to flourish in stable and tranquil times.

The mechanism through which this works is explained – somewhat similar to relative deprivation theory – through the social psychology of the effects of disaster. The psychological stress that accompanies people living through disaster provides the ground on which millenarianism can grow. However, this is not in itself a sufficient condition. There needs to be disorientation, a break-up of traditional groups and what he calls "true society", which leaves people looking for new certainties and landmarks. Barkun thus argues that millenarian movements are "artefacts" of disaster and that "anxiety and meaninglessness create the need for millenarian movements while the dissolution of traditional groups and a state of enhanced suggestibility make membership easy" (pp. 55–56). Millenarianism can then act as an explanation for the difficult circumstances we find ourselves, and it will be able to provide that explanation that will let people make sense of the new world through reference with the familiar.

O'Leary (1994, p. 9) points out, however, that what precisely counts as a disaster is, by Barkun's own admission, "in the eye of the beholder" (Barkun, 1974, p. 60) and even argues, with the example of China's cultural revolution which does not seem have an accompanying disaster (rather, it created the disaster), that "if the world has no disaster to offer, then one must be constructed" (p. 208). Thus the problem with disaster theory – on its own at least – is that it is both rather vague in what counts as disaster, and that sometimes the millennial movement seems to create the disaster rather than vice versa. While it is indeed striking that disaster and millennial movements often tend to go together, it seems that disaster is also not a sufficient reason for millennialism (again, by his own admission; p. 6), but it may not even be a necessary one either. Barkun lists a number of additional factors that might need to be present for millennial movements to arrive and thrive, all of which rather dilutes the explanatory force of disasters for the appearance of millennialism:

> There must be multiple rather than single disasters; a body of ideas or doctrines of a millenarian cast must be readily available; a charismatic figure must be present to shape those doctrines in response to disaster; and the disaster area itself must be relatively homogenous and isolated.
>
> (Barkun, 1974, p. 6)

Rather than disasters being causes of millenarianism, we can instead perhaps claim that millenarianism and disasters reinforce each other: A society in a condition of disaster or even imagined impending disaster may be fertile ground for millenarian movements, while millenarian movements themselves might enact conditions that are disastrous to society (through revolution, terrorism, the Cultural Revolution, Jacobin reign of terror, etc.), or be successful enough in persuading a large enough section of society that disaster is imminent.

Apocalyptic rhetoric: time, charisma and the problem of evil

O'Leary (1994) takes the second and third of Barkun's list of additional factors (that a body of doctrine and charismatic leaders must be present) to indicate that a rhetorical analysis of millenarian argumentation may help us shed light on what is going on. By examining the arguments of millenarianism rather than the social conditions that may or may not cause it, he argues that we can answer different but just as important questions. These he lists as what the persistent appeal of apocalypticism is, what problems it solves, how apocalyptic discourse shapes and is shaped by individual and community worldviews, and what the impact of apocalyptic discourse is on a society's vision of the future (pp. 7–8).

He takes up three particular strands (though O'Leary works with the more technical rhetorical concept of topos) in his theoretical outline of apocalyptic rhetoric: These are evil, time and authority. Turning first to evil, O'Leary argues that apocalypticism helps tackle a problem that in some variation almost all religious systems will have to face sooner or later, the "problem of evil". This is particularly pertinent for the Abrahamic religions, which teach that God is both omnipotent and benevolent. In this monotheistic version the problem of evil is classically formulated by the philosopher David Hume: If God is both omnipotent and benevolent, how can we explain suffering? If suffering exists – which it plainly does – then God can either eliminate it but doesn't (in which case He is omnipotent but not benevolent), or He is willing to eliminate it but can't (in which case He is benevolent but not omnipotent). Either way, He can't be both. While atheists often cite the problem as a clear and final argument against religion, there are several ways of answering it, and indeed a whole branch of theology – theodicy – has devoted itself to it. Whatever the merits of the arguments presented by theodicy, it is at least clear that the problem has been recognised and that theologians believe it can be solved.

Following Nelson Pike (1964), O'Leary argues that there is an assumed premise hidden in the argument that gives rise to the conclusion. The premise is that "an omnipotent, omniscient being would have no morally sufficient reason for allowing instances of suffering" (Pike, quoted by O'Leary, 1994, p. 35). What in this context is morally sufficient? One way of attempting to answer this is St Augustine's argument in *The City of God*: Evil as such is not real since it only exists as "the negation of good". The origin of evil lies in our free will, and free will is a good that a benevolent God has bestowed on us. In Plantinga's words, "to create creatures capable of moral good, therefore he must create creatures capable of moral evil; but he cannot create the possibility of moral evil and at the same time prohibit its actuality" (quoted in O'Leary, 1994, p. 39). Thus the argument is that there is a morally sufficient reason for allowing suffering, and therefore suffering can exist at the same time as a benevolent and omnipotent god.

Showing that evil and moral good can coexist with an omnipotent and benevolent god is easier somewhat with apocalyptic beliefs: If the world does eventually come to an end and souls receive their judgements, evil will be punished, therefore

God can allow evil (and thus also free will) to exist while at the same time being benevolent. If not now, then – in the long run at least – justice is served. And it is the long run, the imagined and narrated future, that the apocalyptic is concerning itself with: "A proposed future good justifies present suffering" as O'Leary concludes (p. 44), a conclusion which he also argues cannot be made with strictly logical criteria but relies on narrative terms, on stories and imaginaries about what will occur in the future.

These arguments, O'Leary adds, are not intended to show that any apocalypticism has been formulated as an explicit response to the problem of evil. However, as part of a wider worldview, apocalyptic religious beliefs skirt around the problem of evil without having to resort to the somewhat complicated and refined arguments of St Augustine (who was no apocalyptic). As a conclusion, O'Leary cites Weber (1963 [1922]) by stating that "apocalyptic eschatology is but one of the many creative forms that evolved from human attempts to resolve the paradox of theodicy that appears in every culture" (O'Leary, 1994, p. 44).

Time is the second major strand that O'Leary picks out. The apocalyptic and the end of the world are concepts that are intimately bound together with our concepts of time, because they concern themselves with our projected ideas of the future (and often a projected continuation and culmination of processes set in motion in the distant past). Referring to the concept of time as "social knowledge", he argues that time (as well as evil) "appear in eschatological discourse as statements about the universe that are assumed to be true" (p. 31). For a specific Christian angle, this social knowledge about time includes that it must have an end. Thus though it is not logically necessary that time has either an end or a beginning, it is the cultural, socially validated knowledge in Christian (and other) societies that it must at some point come to an end. This would be in contrast with other cultural notions of time, for example with Buddhism, which sees it as cyclical. This is a rather brief note on a longer and more sophisticated argument presented by O'Leary, but time as a feature of the apocalyptic will be further elaborated on below.

The third strand is that of authority. As Barkun argued, "a charismatic figure" would need to be present for millennial movements to develop. Charisma and other forms of authority are needed for apocalyptic beliefs (or any other forms of belief) to spread. Theorising authority is one of the main strands of rhetoric: Why do we find a speaker credible? Introducing Aristotle's theory of persuasion, Rapp (2010) notes that

> a speech consists of three things: the speaker, the subject that is treated in the speech, and the listener to whom the speech is addressed (Rhet. I.3, 1358a37ff.). It seems that this is why only three technical means of persuasion are possible: Technical means of persuasion are either (a) in the character of the speaker, or (b) in the emotional state of the hearer, or (c) in the argument (logos) itself.

In order for an argument to be persuasive, therefore, it needs to be credible. As a development of Aristotelian rhetoric, traditionally three types of credibility are distinguished: Source credibility, message credibility and media credibility (see also Rieh and Danielson, 2007). Important here for the purposes of authority is source credibility, which is about the authority of the speaker – this could be their academic credentials, experience or position of authority within society. For example, we would be more persuaded by a speaker who has a PhD on the subject they are talking about or who has personal experience with it.

Within classical sociology of religion, we could argue that source credibility relates to Max Weber's (1963 [1922]) discussion of political (and religious) power. Weber distinguishes between three types of legitimations: Traditional authority, legal/rational authority and charismatic authority. In developing his rhetorical theory of apocalypticism, O'Leary outlines how tradition, rationality and above all charisma have acted as forms of legitimation within Christianity and Christian prophecy. A prophet's legitimacy is derived from charisma in the sense that the audience provides it to them. Without wanting to go into the personal or psychological attributes a person needs to be charismatic (which is a larger and still open question), we can at least say that through having attracted an initial following, the prophet will be able to claim some audience-based legitimacy, which in turn attracts a larger following, through a virtuous feedback loop. This effect is reinforced through the positioning of the prophet's audience as a "universal" audience "a standpoint for judgement that reveals the narrative unity of all space and all time and forecloses every disagreement" (p. 53). By assuming the universal audience, a prophet's charismatic authority has become unquestionable, and "in its initial oracular mode of expression, the authority of the apocalyptic narrative is grounded in the prophet's claim to direct apprehension of the sacred" (p. 53). If authority is granted to the prophet by his audience, then in having the "universal audience" he becomes the ultimate authority. Contested legitimations of authority over scripture, which could previously be held in check by the Catholic Church, were broken with the Reformation and the translation of the Bible, and new prophets emerged that could take authority directly from their claim to be speaking to the divine audience. The history of the church(es) can be recast as a struggle to assert authority and legitimacy over who gets to speak for the universal.

O'Leary states that authority is one of the main areas that need looking at when we want to explain how apocalyptic beliefs spread and evolve, and therefore it threatens to "engulf [his] whole project" (p. 51). Indeed, I would argue that it is one of the main issues that influence not just the study of apocalyptic or religious beliefs, but all beliefs, including scientific ones. Therefore it, along with related issues such as trust, must be a main pillar of any sociology of knowledge. Authority and trust have therefore become an important pair of concepts within the wider sociology of science corpus.

As a sort of conclusion to this however, O'Leary binds his three strands (or topoi) together to inform his approach to the apocalyptic thus:

> When a whole world or cosmos is perceived in terms of the ultimate exigence of evil, and the urgency of this exigence is emphasized by an epochal rhetoric that constitutes time through the imminence of its fulfilment, the theme of

authority becomes central in Christian eschatology. For . . . the apocalypse is (among other things) a mythic narrative about power and authority, and affirmation of divine and spiritual power over and against the idolatrous claims of state authority.

(p. 55)

O'Leary builds the topos of time and chronology into his general theory of apocalyptic rhetoric. From a narrative perspective, the social knowledge that we have through our scriptural understanding of time states that it has a clear beginning and strongly implies that it therefore also must have an end.

> The world is eternal, *or* the world is not eternal. There was a moment of creation, when all things came to be, *or* creation is forever repeated, and all things have always been. Time must have a stop and come to its fruition in a point of ending beyond which is eternity, *or* time is infinite, and will continue on through the unending cycle of eternal return.
>
> (O'Leary, 1994, p. 29, italics in original)

Understanding time also means thinking about the beginning and the end of it. Citing Augustine's arguments on the problem of evil, O'Leary shows that Christian doctrine had to take the option of non-infinite time, because "who could bear an eternity of such misery?" (Augustine, 1972, cited in O'Leary, 1994, p. 30).

The phenomenology of time

It is certainly clear that time and chronology must have a particularly important function in any narrative that involves the apocalyptic, because the apocalyptic is about a particular future event. How we interpret this future event therefore also depends on how we see time itself. For this reason, Hall (2009) takes a historical diversion into various social phenomenological conceptions of time. For concepts as fundamental as this, it is often difficult for us to conceive of other cultural meanings that it can take. However, the strict chronological, linear way of thinking about time that we are familiar with, which divides the years into months, the months into days, the days into hours and so on into ever finer and precisely defined nuggets, is relatively new. Our way of thinking about time, particularly divisions below the "day" unit, only really became possible with the inventions of clocks and calendars. While certain repetitive physical phenomena such as days and years are clearly universal,

> the units of time by which social life is temporally ordered are arbitrary. . . . In some societies in the past, the number of hours of day equalled with the number of hours of night, even though in regions of the world with strong seasons, this meant that in the winter, night hours lasted longer than day hours.
>
> (Hall, 2009, p. 9)

Hall charts various "types of social temporality" (p. 10), which provide different phenomenological accounts of how time is experienced, and outlines how the various experiences of time influence the apocalyptic dynamic of movements and societies.

The "here and now" as an experience of time refers to the everyday lived experiences that are not connected through each other to a wider meaning, characteristic of "primordial" societies. "Synchronic" time, arising out of the structuring of here-and-now experiences in primordial societies, is structured around storytelling, where rituals as collectivised experiences "mark first steps in the institutionalisation of social time" (Hall, 2009, p. 14). Citing the work of Eliade and Durkheim, he remarks that the synchronic time of "elementary societies" is divided between the profane and the ritualistic here and now and describes synchronic time as "organized via the relations between traditions and their re-enactment, which constitute a 'chain of memory'" (p. 15).

Chronological objective time is time as experienced by current, modern society, where clear and objective divisions between years, days, hours and so forth structure our daily lives along with our retrospective experience of history and future events. The shift towards objective time, however, did not occur automatically with the invention and normalisation of calendars and, later, clocks. Instead, the change occurred gradually.

Different experiences and perceptions of time influenced how societies understood history:

> History as an extended chronological sequence of events did not exist. This does not mean that there were no "historical" events. . . . Although history would be invoked, memory of events faded over a short number of generations to the point that events became detached from temporal sequences.
>
> (Hall, 2009, p. 16)

With the slowly changing conceptions of time however, in the Middle Eastern religions, "history could no longer be warded off by the development of cyclical cosmologies . . . the temporal structures of god's purposes became historical, and, in some cases, apocalyptic" (p. 17).

The apocalyptic wave

Landes (2011) builds his theorisation of apocalyptic movements on trying to understand the dynamic development of them, having noted the "semiotic excitement" that seems to push these movements into spectacular growth and just as spectacular contraction, with plenty of violence committed along the way. Landes first characterises two types of people, "owls" and "roosters", as a way of distinguishing between those who are excitedly enthusiastic about apocalyptic prophecy (the crowing roosters) and those who keep calm and realistic (the wise owls). Among the roosters are the charismatic prophets and messiahs but also the charismatic secular leaders with their personality cults. The dynamic rises and falls of millennial movements come about through the interaction between the two. Normally, "most of the time, roosters do not dare crow", and thus "apocalyptic beliefs remain either dormant or concealed" (p. 41); they lie "dormant" as "hidden transcripts" within wider culture but periodically come to the fore as part of protests against established power. The apocalyptic imminence of the end means that the belief that everything will change in the near future and the message of the roosters becomes dominant; "apocalyptic enthusiasts become radically uninhibited" and promise "prophetic vengeance" (p. 43). With the increasing dominance of the roosters' crowing, Landes argues that society moves from "normal time" into "apocalyptic time", when

> they step into a dramatically different universe, one where the final, cosmic revelation, the final resolution of good and evil, now enters the public sphere. Here the "powers" that rule this world will soon receive their just humiliation, if not eternal punishment.
>
> (Landes, 2011, pp. 43–44)

In apocalyptic time, signs and portents are everywhere, and every inconspicuous event can be re-interpreted as something the prophecy has foreseen; people are "semiotically aroused". All this development will of course be continuously opposed by the owls interested in maintaining the status quo (and, of course, their lives). The dynamic of millennial movement, then, depends on the interplay between roosters and owls and the normal and apocalyptic time. Landes contends that this interplay has a regular structure, which he spends most of his book applying to a variety of millennial and apocalyptic movements including the Papuan cargo cults, the French Revolution, UFO cults and contemporary Islamic apocalypticism.

This structure is outlined through another metaphor of the apocalyptic wave. In the first phase of the wave, the "waxing wave", the roosters' crowing starts to build and the movement breaks into public awareness (as to what occasions the initial rise of the wave, Landes points to the disasters and disorientations described in Barkun). In the second phase, the "breaking wave", the roosters' crowing has come to provoke turmoil and briefly become the dominant voice in public life; public life itself is dominated by apocalyptic time. The third phase, the "churning wave", is characterised by the effects of the apocalyptic movement having come to dominate: "The impetus of the movement carries it beyond the failure of expectations" (p. 52). However, the cognitive dissonances (see Chapter 5) between the expectations of the movement and their failure to materialise builds up as the movement spreads; it must adapt and mutate while the owls start finding their voices again. Finally, during the fourth phase, the "receding wave", the owls' voice has become dominant again, society moves back into normal time, and roosters along with their apocalyptic transcript go into hiding again.

Concluding thoughts

The phenomena of millennialism and apocalypticism have been widely documented and described. Though in a narrower sense these terms apply to the messianic, apocalyptic tradition of Judaism and its offshoot religions, the attendant phenomena are possibly features that are more widely shared. These would include charismatic prophets, preoccupation with humanity's future and especially expectation of a catastrophic near future, hopes and dreams of utopian peace to come, also in the near future, and more.

While also offering narratives of hope and the possibility of salvation, peace, and the triumph of good, these movements have become proverbial for violence and violent consequences – inflicted both by the believers on the outsiders and by the outsiders on the believers, but also often the other way around through the persecution of those "dangerous" cults. Much scholarship on millennial movements has focussed on the violence and other possibly nasty consequences either to explain it, to avert it or to persuade otherwise (Barkun, 1996; Wessinger, 2000; Juergensmeyer, 2017).

The sociological explorations of millennialism outlined here have focussed on trying to explain what makes them so attractive, what wider social circumstances a society might be going through in order to make their appearance more likely, to explore their possibly different conceptions of time, their "semiotic excitement". All these have their virtues, and as I explained above, I don't think they are either necessarily in competition with each other, or wrong – rather, together they start building a wider sociological picture that might indeed help us understand them a bit better.

As several of the commenters introduced here have noted, there are many parallels between the movements outlined in this chapter and the movements that worry about environmental or climate collapse. Globus and Taylor (2011) discuss them as "environmental millennialism", even though they note that millennialism is clearly not a "native category" by which they would describe themselves. Wojcik (1997) categorises them with the avertive types of millennialism. Maybe the obvious difference is that environmentalism is something we should actually be worried about, since the science is conclusive that climate change is happening, and that it is going to be a civilisation-threatening risk. But then, to those living within any other millennial movement, the threats won't appear any less real either.

The approaches outlined here have so far not really addressed the knowledge itself, how it is formed, how people interact with it and believe in it, where it comes from and what is considered reliable. However, one approach to the apocalyptic that does address knowledge is Festinger's cognitive dissonance (Chapter 5), however while widely influential as a social psychology of millennial movements, as far as I can tell this has not been further developed even though within social psychology new approaches and theories have been developed that could enhance our understanding of millennialism.

In order therefore to understand current "scientific" millennialism and its relation to the apocalyptic better, I am interested in picking this up and looking further into the nature of the knowledge itself and how our cultural expectations honed from centuries of millennialism have shaped how we make sense of the new scientific knowledge. This, I believe, might be where my perspectives from the philosophy and sociology of science can help in adding to our social theory of the apocalyptic. In the following chapter, therefore, I will try to connect these cultural expectations and representations of the end to our current expectations of what the end might be with the help of social representation theory.

Chapter 4

Apocalyptic visions

Introduction

In much of the sociological literature reviewed in the previous chapter, the end of the world is usually a fairly undefined event, with the spiritual, psychological and sociological implications, the imaginaries about utopian millennia, the role of saviours and messiahs, good and evil, redemption and judgement all usually deemed more important than what actually happens to bring it about – or indeed, what it is precisely that is being brought about in the first place and who is doing it. This is partly a perfectly reasonable result of the need to find generalisations of course, social explanations for millennial movements that bind these various imaginaries together, regardless of what they believe will happen precisely, and the tales of post-apocalyptic utopias, battles between good and evil and so on are indeed in a way more important to the apocalyptic narrative than the method through which the end is achieved. However, the variety of imagined ends and how they connect with each other and over time between various end-time movements and current, secular expectations is instructive, and it is not merely the manner in which the world is supposed to end but also who bears the responsibility for it and how severe it is going to be.

The interpretative flexibility about what we mean by the end of the world is arguably one of the reasons it is such a powerful concept, as people can project their own current fears for the future onto a host of different and often contradictory interpretations and expectations derived from our shared cultural knowledge of the apocalyptic. It also enables failed predictions of the end to be fairly easily explained through explanatory shifts in what the end really means. This, however, might set a trap for the social scientist, because when *we* write about the end of the world, our main subject is slippery and unclear, and theories we build to deal with one type of end can be too easily transposed to other types without clear thought on whether the central insights are still applicable. Linking millennialism as a sociological concept with fears about climate change or environmental degradation, as Globus and Taylor (2011) do, or I am about to attempt later, might be one such trap. We may see parallels in how people react and how movements form and develop, but are we really talking about the same things? In the "scientific" apocalypses, the means of the end is often deemed the more important thing to know than the nature of evil or redemption, even though I will try to argue that all these are important parts of how we as a society tend to think about climate change or the AI apocalypse, or meteorite impacts as well, they are just better hidden.

Thus while this book is also about the sociological implications of the end of the world more widely rather than about what will actually happen as such, an overview of the actual visions for the end and how they interrelate will be interesting. One of my underlying arguments will be that our social representations and social knowledges about the world are interconnected, and as such we are inclined to see parallels for example between biblical floods and the rising water levels of climate change – particularly so when we read evangelical end-times preachers such as Hal Lindsey (1970) directly comparing the effects of nuclear weapons with the fire and brimstone of Revelation, thus making a case for nuclear war having been prophesised in scripture. This to me at least shows that what we imagine the end of the world to be has clear sociological implications. This chapter will look at three aspects of how the world will end in turn – the severity of the destruction, the nature of the destruction, and the agent of the destruction – before providing a brief outline of social representation theory as a conceptual tool for linking all these different apocalyptic imaginaries together.

Severity of the destruction: apocalypse and risk

One of the ways in which the end of the world can vary is in the consideration of *what* will actually end. We could, and often do, talk about either the end of everything, the end of life, the end of humanity, the (mere) end of civilisation or only civilisation *as we know it*. As a way of cataloguing potential apocalypses along this line, the science fiction author Isaac Asimov (1979) organised world-ending catastrophes into a fivefold hierarchy. This was done in a sense more as a structure to build a popular science book around, and as such he concentrated on then-current scientific explanations of what might happen rather than incorporating spiritual or religious ideas on the end (apart from a short introduction). However, as a starting point this might be useful. Asimov's first-level apocalypse is the end of the universe and everything in it; in this context, he writes for example about the heat death of the universe. The second level is illustrated by the destruction of the sun, for example its eventual demise or the possibility of being swallowed by a black hole. The third level is the destruction of earth (e.g. through an asteroid impact). The fourth level is the destruction of life. The fifth is the destruction of civilisation.

One of the interesting features of Asimov's book, compared to similar, more recent books, is his optimism. Most of his explorations on the possible ways the world can end at the lower levels of his categorisation (i.e. destruction of the universe, the world or life on earth) have the reassuring tone of the scientist explaining how unlikely or how far into the future the catastrophic event is. It is also clear that some of the catastrophes Asimov writes about are not quite the apocalyptic ends I'm interested in; the heat death of the universe is too far removed in time to worry us, similar so the end of our sun's natural lifetime.

Some possibly more worrying catastrophes, such as meteor impacts, are for Asimov not much to worry about because they happen so rarely that, by the time they are likely to happen, humanity will have advanced enough to be able to detect and deflect meteorites or to decamp itself onto new planets. The catastrophes that do worry him are less destructive in that they (merely) threaten civilisation or at least a large chunk of it. The risks then more or less balance out and the more likely and imminent a catastrophe is, the less catastrophic its consequences.

Asimov's optimism stands in contrast to the predictions of doom that Karplus (1992) wrote about. Karplus noted that he was irritated by the proliferation of scientific and popular scientific doomsday projections he was seeing in the bookstores, such as the AIDS epidemic, acid rain, the ozone layer, overpopulation, economic collapse, earthquakes and climate change. He intended his contribution to be a sober analysis of these varieties of scenarios, some of which he concluded are indeed worth worrying about, others maybe less so. A persistent theme in his analyses, however, was the opportunity costs associated with responding to these threats; for example throwing too much money and effort into combating one particularly high-profile infectious disease outbreak may draw attention away from combatting other diseases or neglecting wider public health, thus saving fewer lives in the long run. Karplus' catastrophes were a mix of civilisation ending and "merely" disastrous; as he outlined himself, a topic such as earthquakes are much more localised and thus less worrying for civilisation as a whole than a pandemic or climate change would be.

Jared Diamond's (2011) best-selling analysis of the environmental and social factors in the collapse of historic civilisations focussed likewise on the "mere" collapse of civilisation rather than any of the "higher" apocalypses – though of course the civilisational collapses Diamond wrote about are all very dramatic and apocalyptic for those who lived through them.

I have briefly noted a few science books written for a broadly popular audience above, as an indicator of how public debates have been surrounding the apocalypse from a more or less scientific, secular perspective. As different as they are, these books are part of a genre that several other sociological writers on the apocalyptic have also spotted (see the more extensive lists in Wagar, 1982, p. 4 and fn. 4, and Wojcik, 1997, p. 98). These are books published in the post-war period, possibly influenced by the technologically apocalyptic societal introspection accompanied by the cold war, which examined the scientific and philosophical basis behind some of the ways the world might end. These explorations can be optimistic like Asimov, an attempt at unbiased realism like Karplus or pretty depressing like Diamond, but they all connected, and they differ from previous apocalyptic work in that they are usually bracketing out the religious element completely, even if it is used sometimes in introductory chapters to set the scene.

An interesting new departure in this genre was a book by the philosopher John Lesley (1996), which drew attention to philosophical and ethical as well as scientific issues, in particular an adaptation of Brandon Carter's anthropic principle (Carter, 1983) based on the probabilistic reasoning that we are most likely to be among the most numerous generations in history. If, due to the recent population explosion, say, 70% of all humans that ever lived are alive now, then it is most likely that the philosopher thinking about this is of them. Alternatively, imagine humanity will go on to live for millennia into the future, colonising the galaxy and so forth; then the chances that we are born as part of the >1% of humanity that lived at the beginning of humanity is very low. This implies that population numbers are likely to be much lower in the future, and some sort of near-future catastrophe is likely (Lesley argues that while this argument appears very unintuitive, it is surprisingly difficult to refute).

The philosophical turn has in turn partly inspired the philosopher Nick Bostrom to turn his attention to "existential risks" (2002, 2013; Ord, 2020), and to gather colleagues together in the development of this area as a wider interdisciplinary academic field. This has led, among other things, to the establishment of dedicated centres in both Oxford and Cambridge University, the Future of Humanity Institute (FHI) and the Centre for the Study of Existential Risk (CSER), respectively. The CSER sees existential risk as the risks of "human extinction or civilizational collapse" (CSER, 2020); this then does cover the apocalyptic conceived very broadly, having been co-founded by the Cambridge astronomer Martin Rees (himself also a contributor to the apocalyptic popular science genre; Rees, 2004).

The concept of risk is usually (but not always, especially in sociology) described as a measure that combines the probability of an event happening with the severity of the consequences of the event. Thus a risk with low probability but severe consequences is similar in magnitude to a risk with high probability but relatively benign consequences. As originally applied to financial risk, the severity of the consequence was relatively clearly defined; this is rather less clear when we the consequences are of a different nature. Nevertheless, the risk of human extinction, if not civilisational collapse, can be thought of as the ultimate severe consequence, and therefore existential risk is in a way a clearly apocalyptic concept.

The choice of language here is interesting, however. Talk about risk, rather than apocalypse, end of the world, or a similarly emotive and historically loaded term, has wider connotations that a serious university research centre may want to avoid (unless of course, it is studying the historical notion of the apocalypse directly). Instead a discourse on risk rhetorically elevates the topic to the scientific, lending academic legitimacy to a topic that could otherwise be dismissed as apocalyptic fearmongering. And yet, in almost all of this work, mention of apocalypse or doomsday is made frequently, usually in the scene-setting of the introduction but not confined to it. With this observation I do not mean to assert that risk – or even existential risk – is merely a 21st-century reformulation of the apocalypse, however I intend to make the argument that, like God and Nature, existential risk and apocalypse fulfil some of the same explanatory and rhetorical functions in wider societal discourses on impending catastrophes (i.e. they perform similar discursive work).

Maybe in contrast to the current fears on apocalypse covered by the existential risk remit, the question of the severity of the cataclysm is more nuanced than more directly apocalyptic discourse in a traditional or religious sense. The contemporary popular and academic work I have outlined here, after all, has made clear distinctions of the severity of the event and ordered the analytical response accordingly. Biblical apocalypse, by contrast, posits the end of the (immanent) world and humanity and distinctions between end of the universe, the physical end of the earth or solar system, or the mere end of life on earth (as Asimov outlines it) are in a sense fairly irrelevant here; the importance lies with the millennium, judgement or salvation.

However, there are nuances in the severity of the destruction more widely than in Asimov's or the existential risk scheme if we consider the spiritual element – clearly if we believe in life after death, and indeed salvation and redemption after death, then death itself is not as severe an adverse event after all, and the severity then depends on the theological expectations of what awaits us afterwards. Apocalypses can either be entirely nihilistic, in that they wipe out all human existence with no afterlife to come, or it can wipe out merely the immanent world and our souls continue meeting our various fates, in paradise or elsewhere.

In this sense we can maybe discern a clear difference between the higher-level apocalypses (in Asimov's scheme), which are the real, proper and nihilistic end of everything or at least of humanity, and biblical apocalypses that are instead a stage in the development of the history our immortal souls. However, this division is not very clear-cut, just as the division between a secular and a religious worldview is also problematic. And as the contributors to Wessinger's (2011b) collection have shown (as well as the wider area of cross-cultural studies of millennialisms), millennial beliefs about the post-apocalyptic world vary widely in any case. Ideas that derive from our continued existence after apocalypse have lived on and acquired new meanings in the secular apocalypse, particularly so when the destruction is not that of the whole world but merely that of civilisation or most of humanity, in which case a millennial kingdom can spring up from the ashes of the catastrophe and offer us salvation or redemption that way. In the post-millennial (or, after Wessinger, the progressive millennial) version of millennialism, it is the optimistic prospect of human progression and ability to form a better society that sustains the hopes of millennial utopia, and this is directly comparable to the enlightened post-environmental societies ("ecotopias") envisaged by some contemporary environmental groups (Garforth, 2018).

Nature of the destruction

With maybe the exception of the higher-order apocalypses that Asimov introduces at the beginning of his book (supernovas, heat death of the universe, etc.), the popular science apocalypse genre often features a very familiar roll call of possible events that might bring about the end, however "end" is defined exactly. These include conventional and nuclear war, famines, infectious disease outbreaks (either natural or as biological warfare or terrorism), floods, earthquakes and so on. Most of these are also in some form present in the scriptural description of the end that forms part of our cultural heritage. But generally, even novel risks, such as climate change, are seen as threatening because its various consequences – sea-level rise, migration crisis and accompanying resource wars, famines due to change in weather patterns – are all types of disaster listed in scripture and thus culturally familiar.

One of the most striking and often visualised elements of the visions of Revelation is that of the four horsemen, appearing in many interpretations as anthropomorphisations of disaster:

> I watched as the Lamb opened the first of the seven seals. Then I heard one of the four living creatures say in a voice like thunder, "Come!" I looked, and there before me was a white horse! Its rider held a bow, and he was given a crown, and he rode out as a conqueror bent on conquest. When the Lamb opened the second seal, I heard the second living creature say, "Come!" Then another horse came out, a fiery red one. Its rider was given power to take peace from the earth and to make people kill each other. To him was given a large sword. When the Lamb opened the third seal, I heard the third living creature say, "Come!" I looked, and there before me was a black horse! Its rider was holding a pair of scales in his hand. Then I heard what sounded like a voice among the four living creatures, saying, "Two pounds of wheat for a day's wages, and six pounds of barley for a day's wages, and do not damage the oil and the wine!" When the Lamb opened the fourth seal, I heard the voice of the fourth living creature say, "Come!" I looked, and there before me was a pale horse! Its rider was named Death, and Hades was following close behind him. They were given power over a fourth of the earth to kill by sword, famine and plague, and by the wild beasts of the earth.
>
> (Revelation 6:1–8, KJV)

The conventional identification of the horsemen as anthromorphisations of war, famine, disease and death (though there is some debate about these interpretations; Osborne, 2002) already incorporates many of the recurring themes on how the world will end; with the exception of death, all of these represent catastrophes which either by themselves or in combination can cause the apocalypse. These are of course not exhaustive (the flood, another recurring theme in how the world will end, is not represented here), but they make for a good starting point in outlining the various possible catastrophes.

War

Various apocalyptic accounts in the Bible indicate war as one of the events during the end time, just as it is one of the "signs of the hour" in popular Islamic eschatology. Hordes from the east will invade, forces from the south, Gog and Magog from the north, there will be a decisive battle at Armageddon and so on.

Despite war being one of the predominant modes of destruction, in a way the idea that warfare is an integrative part of how the world will end is theologically somewhat problematic, because the war is both prophesised and thus preordained but also waged by human beings who presumably have agency and the potential to refuse to do evil. If the hordes of Gog and Magog are needed to fulfil God's plan, how can their actions be evil? This question has been the matter of some substantial theological introspection. This is not the place to add to or critique these discussions, but it is probably worth pointing out that there has been a shift in the perception of who is to blame, for war has probably been at work here as well. To the ordinary person caught up in a war, it can indeed feel like divine punishment rather than the outcome of deliberate political decisions.

Anxieties about war as a potential civilisation-ending event have clearly not lessened through secularisation of our imaginaries of the end. The traumatic worldwide wars of the early and mid-20th century have shown that they have the potential to be civilisation ending, especially if we take the development of weapons of mass destruction, in particular nuclear weapons, into consideration. These, and in particular nuclear weapons, have given rise to apocalyptic discourses in their own right.

However even conventional warfare has the potential to become apocalyptic, in that it can cause economic collapse, famines, diseases and of course direct loss of much human life. The prospects of war and armed conflict are often raised as one of the secondary consequences of more complex catastrophes. Severe climate change, for instance, will have consequences on the arability of already marginal farmland in developing countries, leading in turn to famines, migrant crises and the attendant potential armed conflicts over increasingly scarce farming and food sources.

Famine

Like war, although given its own horseman of the apocalypse in popular culture, famine itself is more often an accompanying disaster of a larger catastrophe. The disruption of warfare for example leads to untended fields due to conscription of the workforce, or in a nuclear war scenario it might lead to contaminated crops. Conversely, climate change will lead to decreasing arability of farmland, which as noted above might in turn lead to armed conflict. An infectious disease outbreak, either natural or through biological warfare, might decimate crops and livestock and thus lead to famine.

Apocalyptic science fiction novels and movies, such as Stephen Baxter's recent novels *Flood* and *Ark* (2008, 2009), often present the apocalyptic consequences of famine caused itself by worldwide floods engulfing the planet. Similarly, famine is a recurring theme in post-apocalyptic dystopian fiction, such as the *Mad Max* movies, where food (as well as fuel) scarcity is one of the narrative driving forces (Hassler-Forest, 2017). However, the ultimate cause of the food scarcity is usually clearly marked as something else: An unspecified, probably nuclear war for *Mad Max*, and floods for Baxter.

One particular element of environmental degradation that has by itself been pointed to as a potential cause of a worldwide famine in contemporary news and popular scientific discourse – where the famine element appears to me at least as the main driver of the apocalypse rather than being a secondary effect – is the current discussion of insect disappearance. While itself caused by a conflagration of various issues such as habitat loss (through climate change, destructions of larger ecosystems for farmland), excessive insecticide use, mysterious diseases, or disruptions in the food chain through other causes, the disappearance of insects has alarming consequences for food crops that rely on insects for pollination.

Pestilence

Another horseman of the apocalypse, disease outbreaks are justifiably feared as catastrophes that could bring about the end of civilisation. Like war, disease has featured as a catastrophe in various apocalyptic visions from the beginning. In terms of sheer impact, a virulent infectious disease such as the Spanish flu epidemic of 1918/1919 can have a devastating impact, and as the coronavirus/COVID-19 pandemic at the time of writing showed, the disruptions they can cause are devastating and global and evoke apocalyptic imagery in popular discourse (Snowden, 2019), even if on sober analysis a disease like COVID-19 is hardly going to mean the end of civilisation.

Like war, fears about disease outbreaks have been very justified over the centuries, and even if they have not yet brought about the end of humanity or civilisation, they have had a major impact: The Black Death in the 14th century killed off a large amount of the population and fundamentally altered the political economy of Western Europe (Cohn, 2007). Western infectious diseases have played a major role in the European conquest of the Americas through the importation of diseases to which native populations had no immunity, and thus they can be viewed as properly apocalyptic for at least some civilisations (Diamond, 1997).

Disease outbreaks have been widely seen as a punishment from God or the gods, and before the development of epidemiology, lack of detailed understanding of how they spread would have heightened the feeling that they are capricious and unpredictable (and, as COVID-19 has shown, even modern epidemiology does not eliminate this feeling completely).

Karplus (1992) noted the AIDS epidemic as one of the eight catastrophes he outlined in his book that he feared the most, and while HIV/AIDS has thankfully not developed as the world-threatening disease it might have appeared to become in the early 1990s, as COVID-19 demonstrates diseases can come completely unannounced and out of the blue. However, though Ord (2020) included pandemics as apocalyptic risks, he argued that it would be a genetically engineered disease (e.g. from bioweapons development) that posed a far higher risk than a naturally occurring disease.

In fiction, the disease apocalypse has featured prominently. Kim Stanley Robinson's *The Years of Rice and Salt* (2003), for example, set up a post-apocalyptic

Europe that imagined a black death having wiped out all of the (European) population. *The Plague*, the existentialist novel of Albert Camus (1972 [1948]), imagines the apocalyptic outbreak of a Black Death–type disease in 20th-century French Algerian Ouran (see Kermode's 1967 discussion on Camus). Finally, the zombie apocalypse, a relatively recent narrative (Webb and Byrnand, 2008; Bishop, 2009), also imagines the zombification mechanism at the heart of the story to be a disease – even if not always particularly spelt out as such, the mechanism of transmission of zombification is modelled on a folk epidemiological understanding of infectious disease propagation.

Floods

The story of Noah and the flood is one of the most famous Bible stories even if not considered an apocalyptic event, as it outlines events from the distant past rather than the imagined near future. Nevertheless, as a bringer of worldwide humanity- or civilisation-ending catastrophe, it features heavily in other biblically derived apocalyptic visions and elsewhere. Like war, famine or disease, a worldwide flood is not unique to Christian or Judaic religious traditions. Floods have been part of folkloric tradition throughout the world (Dundes, 1988), and the flood continues to be a recurring motif for apocalyptic and millennial narratives as one of the ways the world may come to an end, for example the UFO cult famously studied by Festinger, Riecken and Schachter (1956, see Chapter 5).

The flood has also become one of the main narratives through which the potential devastation of climate change has been communicated. While other consequences of global warming may be just as destructive, or more so, it often is the rising sea levels that are mobilised to visualise the direct impact.

Finally, floods appear frequently in apocalyptic and post-apocalyptic fiction. I have already introduced Stephen Baxter's work above as an example when discussing famine; Baxter's novel envisages a large, sustained (and mostly unexplained) sea-level rise that eventually floods even the highest mountains. A similar scenario is explored in the movie *Waterworld*, set in a post-apocalyptic future where most of the planet apart from some of the highest mountain peaks is flooded.

Fire, brimstone and the stars falling from heaven

Fire, one of the most fundamental forms of destruction, features heavily in biblical and other apocalyptic accounts, often as a companion to the wider destruction of war and as punishment for the wicked. In Ezekiel's apocalyptic account, for example, it is fire that devastates the invading armies of Magog: "And I will send a fire on Magog, and among them that dwell carelessly in the isles: and they shall know that I am the Lord" (Ezekiel 39:6, KJV). And again, in Revelation:

> The same shall drink of the wine of the wrath of God, which is poured out without mixture into the cup of his indignation; and he shall be tormented

with fire and brimstone in the presence of the holy angels, and in the presence of the Lamb.

(Revelation 14:10, KJV)

As much as fire is more an accompaniment to the wider apocalyptic narrative of destruction than a direct cause of the destruction, and didn't deserve its own horseman in popular imagery, the phrase "fire and brimstone" has nevertheless become proverbial as a popular indicator of God's wrath.

As a "scientific" apocalypse, the connection with brimstone points to the possible narrative connections with volcanoes and supervolcanoes that writers such as Asimov and others have explored as world-ending catastrophes; stretching the connection maybe a bit too far, it also connects to the possible apocalyptic consequences of a nearby supernova.

It is, however, the spectre of nuclear and thermonuclear warfare that caught contemporary imaginations as to the possible connection of biblical prophecy with 20th-century signs of the hour popular in Christian evangelical writing. Hal Lindsey (1970), for example, makes the connection explicit in equating the fire that rains down in Revelation with thermonuclear warfare (see Lahr, 2007). Following this connection, the contemporary fictional treatment of the fire and brimstone motif in the form of nuclear apocalyptic fiction and science fiction is, of course, extensive (Shapiro, 2002).

Then finally, probably not worth its own subsection, a contemporary apocalyptic scenario that is much discussed in the popular science genre and well represented in disaster fiction (e.g. *Deep Impact* and *Armageddon*) is that of meteorite impacts. While not particularly as prominent in scripture as war or disease, this apocalypse however has a potential biblical counterpart: Revelation 9:1 (KJV) talks about stars falling from heaven.

By making all these connections I may very well be guilty of the same "semiotic excitement" (in Landes' 2011 phrase) experienced by prophets interpreting scripture with an expressed desire to see equivalent signs in the contemporary world rather than finding real narrative connective threads – this is a possibility I feel more keenly in this subsection than in the ones on war or disease. However, the fact that these connections can be made so easily, not just by me but also by earnest biblical prophecy interpreters such as Hal Lindsey, shows at least the imaginative ease through which contemporary, secular fears can be understood through a biblical lens.

Human hubris as the force of destruction: miscellaneous other catastrophes

These biblical motifs of the modes of destruction listed above, however, do not exhaust contemporary apocalyptic imaginaries about how the world might end. On the possibly more whimsical end of the scale of apocalypse that does not appear to have a biblical equivalent, the years approaching the turn of the millennium in

2000 have featured the much-discussed technological problem of the Millennium Bug as a possible apocalyptic scenario (McMinn, 2001). Other discussed apocalypses, of lesser perceived probability but of larger potential destructive impact, are scientific risks such as the ignition of the atmosphere through nuclear weapons. As Ord (2020) discusses, with current physical knowledge this fear turned out to be unfounded, yet in 1945, when the first nuclear weapon was fired, this was considered a distinct if unlikely possibility – and yet the Trinity nuclear weapons test still went ahead, initiating hair-raising ethical issues. Similarly, there were fears over high-energy nuclear science experiments at the Large Hadron Collider (see also Lesley, 1996), which are unfounded but still raised plenty of apocalyptic popular discussions (Kahn, 2008). Advances in technology and scientists' propensity to go ahead even when there are small chances of resulting doomsday (as evidenced by the Trinity test) has given rise to a wide variety of other technological apocalyptic scenarios, such as the "grey goo" of nanotechnology (Anderson et al., 2005). The prospect of artificial intelligence and its apocalyptic potential is the one scenario that worries some of the existential risk researchers such as Ord (2020) and Bostrom (2014) the most; this is also a consequence of technological progress.

While these technological risks are all novel in their own way, a connecting thread is that of human technological hubris coming back to haunt its makers. Seen through this angle, these apocalypses are indeed linked to older myths from our cultural heritage, that is the Promethean (Blumenberg, 1979) or Frankenstein (Turney, 1998) myth. This complex also links back not just to the ancient Greek myth of Prometheus, who stole fire from the gods to give to humans, but also wider biblical narratives of the dangers of forbidden knowledge and original sin. The element of technological hubris as the cause of our destruction also connects the Promethean myth to the other main technological apocalypses such as climate change, environmental degradation or nuclear warfare, whose consequences (in terms of floods, famine or fire raining down from heaven) have been discussed above. As Turney (1998) shows, the Frankenstein myth has become a common thread in fictional and especially science fictional treatments of the apocalypse.

The narrative of human hubris as the force that will destroy us links to the final aspect of the apocalypse to be discussed in this chapter: The question becomes not so much what will destroy us but who is responsible for the destruction.

Agent of the destruction

Wojcik (1997) argues that apocalyptic visions have darkened as they changed from the transformative millennialist hopes of the revolutionary movements to the secular, nihilist, and scientific visions of destruction of the second half of the 20th century:

> Science the end of World War II, visions and beliefs about the end of the world appear to have become increasingly pessimistic, stressing catastrophic

disaster as much as previous millenarian visions emphasized the imminent arrival of a redemptive new era.

(Wojcik, 1997, p. 98)

The perception that the scientific, secular catastrophes are essentially nihilistic, representing an end of the world without redemption or any wider deeper meaning is widespread, and to a large extent this is justifiable. However, the loss of meaning is by no means complete and can be easily reinserted in the secular apocalypse. As I noted earlier, meaning is often better hidden within the narratives of secular apocalypse, but nevertheless it is there if we look more closely. One way of seeing this is taking the nihilistic element seriously and pursuing the ethical questions that arise from that. The labelling of climate change as an "apocalyptic" in popular discourse, and thus linking it to the fatalistic expectation the word evokes also acts as a smokescreen, as Skrimshire (2010) argues, suggesting that "the fight is over", and this fatalism is in itself an ethical position to take. Nihilism is in a way a self-contradictory position to take generally, as the denial of meaning is itself a significant and signifying act (Diken, 2009).

But linking secular catastrophes directly to scripture is another clear way through which we can find meaning in the end, and re-insert transformative, millennial hopes back into the apocalypse; natural disasters and ecological catastrophes might after all be one of the many ways through which God might choose to bring about the end. The cold war, for example, has been a major theme in post-war US evangelical premillennial theology (Lahr, 2007); thermonuclear war may after all be the fire and brimstone that signals to us that the millennial kingdom is close.

While this is an important strand in the wider religious interpretation of what various secularised catastrophes really mean, the constructions of meaning go wider than the explicitly apocalyptic interpretations of US evangelists. Sociologists and anthropologists such as Latour (2017) and Danowski and Viveiros de Castro (2017) argue about the deep connections between religious and environmental meaning making on the ecological and climate crisis. As I will outline further in Chapter 9, the connections between environmentalism, spirituality and even organised religion run deep.

Meaning can be inserted into the catastrophe without drawing either on explicitly religious traditions or nihilistic fatalism. Rohloff (2012, 2019), for example, outlined the moralising of environmental discourse which she examined as a (justified) moral panic, in a development of Cohen's (1972) classic concept, and which itself is part a wider civilising process as described by Elias (1994). This moral dimension to the environmental crisis is clearly related to its agency, who brought about the crisis and who is responsible for acting on it. While potentially the finger can be pointed at individual actors, the environmental crisis is more often interpreted as a collective human responsibility; the way we live and the way we consume collectively has an impact on the severity or occurrence of the coming catastrophe. Thus environmental morality is bound up with the ethics of individual virtues, and parallels can be drawn between the individual redemption narratives in (Protestant) religious morality. Humanity can be redeemed through the individual leading a virtuous life, just as the environment can be saved through individuals adopting an environmentally conscious and carbon-low lifestyle. Rohloff, in comparison with other lifestyle- and consumption-related moral panics, terms this the "carbon temperance" movement.

In all of these, a meaning is found within the apocalypse that looks towards the actors responsible for it, and where and how we find redemption for our actions. It is agency and responsibility for the end, then, that I would suggest provides an interesting third dimension to how visions of the apocalypse have been understood and made sense of.

The agent of who or what brings about the catastrophe can be divided maybe into the divine (God sends the wars, plagues, etc.), we humans ourselves (man-made climate change, nuclear war, etc.) and finally everything else (e.g. alien invasions, meteor strikes or other natural catastrophes). It is interesting, then, that in many accounts of world-ending catastrophe the responsibility usually does come back to humanity. This is obviously so in the catastrophes like climate change that are caused by us directly and that lead us back to the clear morality tale of the Prometheus myth. But it is just as clearly our responsibility if God in Her anger at our amoral ways chooses to end the world, or even if She destroys it for our own moral good, to save us from ourselves even if there is no anger involved.

Even completely natural disasters caused by neither human technology nor the wrath of God point towards often clear ideas of who is responsible, if not for the disaster itself then for the suffering it has caused. Infectious disease outbreaks, even though natural phenomena, can and routinely are blamed on dirty habits of foreigners, as the early stages of the COVID-19 pandemic demonstrated. The devastation caused by earthquakes can be blamed on corruption in the building industry or complacent monitoring by scientists, as was the case in the 2009 L'Aquila earthquake in Italy (Prats, 2012; for the wider point, see Beck, 2007, and the discussion in Chapter 6).

Reaching into the past – myths and social representations of apocalypse

The change in our retelling of imminent catastrophe, from God-induced but ultimately our fault and responsibility to the secular catastrophes that are directly both our fault and therefore our responsibility, is therefore not as stark as it might seem at first sight. The persistence of the Promethean myths on the dangers of human knowledge and technology throughout the age of secularisation is one important demonstration of this. Using Prometheus as one of his examples, the German philosopher Hans Blumenberg (1979) examined the persistence of myth in human culture (as a counterpoint to the view of a progressive enlightenment advanced by Adorno and Horkheimer, 1997 [1947]), arguing that myths fulfil an important function as the narratives we use to make sense of our lives and the world around us.

The apocalypse, then, even clearly secular and or fictional apocalypse, is one of the persistent myths or narratives that structure how we think about the world we live in, and which can be infused with a wider meaning that is itself derived from our background cultural and social knowledge. I would argue that using these myths and ways of imparting meaning onto a secular and atheist world is not merely reinventing religion or a psychologically comforting way of escaping nihilist despair at the end of our world. Instead, finding meaning derived from myth can be a sensible way of reacting to and even mitigating events, a heuristic through which we can understand what is at stake and undertake appropriate action.

Our collective imagination – as individuals, as identity groups and as larger societies – about the future and imminent catastrophes shapes, and is shaped by, our wider social narratives that inform our day-to-day imaginaries. Our cultural upbringing within which certain ways of thinking and understanding the world are firmly established colours how we interpret new messages and how we try to make sense of new information, new situations and new concepts.

The social psychologist Serge Moscovici (2001) has introduced the concept of social representations as a general social psychological theory to explain how people make sense of new concepts: As we have all our preconceived ideas and understandings of how things work, any new concept or idea is *anchored* conceptually to things we have previously understood, and this anchoring process can differ depending on the social context in which it happens, and it consequently induces different attitudes towards the new concept. Electronic cigarettes, for example, can be anchored to conventional cigarettes, colouring them with the generally negative attitudes we have towards smoking, or they can be anchored to therapeutic nicotine, imbibing them with the generally positive associations we have towards stop-smoking therapies (on this conceptual ambiguity of electronic cigarettes, see Tamimi, 2018). Either way, the alien and strange therefore is made familiar and intuitive. It is thus both the meaning of a concept and how it is produced and transformed, but also the social context in which this happens, that influences our representation (Jovchelovitch, 2007, p. 13).

Through anchoring, the unfamiliar is being made familiar but can also include more abstract and therefore more nebulous and insubstantial concepts. *Objectification* is the process that gives new and unfamiliar abstract concepts a seeming reality through anchoring to more concrete concepts (similar to the concept of reification). Moscovici (2001) introduces the concept of objectification through the development of scientific theories, which start out as abstract theorisations but on becoming more accepted and taken seriously they "turn out to be normal, credible and brimful of reality" (p. 49). Moscovici describes personification as a form of objectification where a new unfamiliar and abstract concept is being made familiar through reference to a person (or the idea of a person). This is the mechanism that underlines the anthropomorphistic tendencies all cultures seem to have which imagine forces of nature and other abstract concepts as persons, for example the personifications of nations (e.g. Britannia or Uncle Sam), races or other abstract concepts such as Justice or Liberty. Similarly, ancient gods personified the natural forces they govern in a process similar to modern-day hero worship and personality cults. A personification is "a social representation that transmutes words into flesh, ideas into natural powers, nations or human languages into a language of things" (p. 54). In this vein, the four horsemen of the apocalypse, for example, personify the abstract concepts of destruction through various means.

Noting social representations as an approach of particular value to research on the public understanding of science – because it deals with the dynamics and social contexts through which knowledge travels between communities, in this case "public" and "science" – Bauer and Gaskell (1999) pick up Moscovici's 1961 doctoral dissertation (republished as Moscovici, 2008) as the main example. Moscovici describes how different sections of French society understood the concept of psychoanalysis, with the different groups bringing their own different social contexts and background knowledges into how the concept was represented· and communicated, influencing the contents (e.g. the anchoring, objectification), process (e.g. propaganda, diffusion) and consequences (e.g. stereotypes, attitudes) of the communication. This integration was labelled as "the masterstroke" of Moscovici's analysis by Bauer and Gaskell (1999).

The mere act of making the unfamiliar familiar through seeking parallels with already familiar concepts – which often is the only way we can get to grips with new phenomena – explains a certain amount of conservativeness into our thought. Indeed, Moscovici thought that his theory of social representations neatly explains the famous observations made by the philosopher and historian of science, Thomas Kuhn (1962), that science progresses through a series of relatively rare revolutionary paradigm shifts. Between these shifts, any newly encountered phenomena (including ones that would appear to directly contradict the prevailing theoretical framework of that science) will be explained in terms of conventional scientific understanding within the prevailing paradigm (Moscovici, 2001, p. 151; Sammut et al., 2015).

Moscovici's own elaborations of social representation theory posits social representations as relatively static. Developing social representations theory further in order to account for more dynamic shifts over time, Bauer and Gaskell (1999, 2008) propose that we should see a social representation as the triangular relationship between the object of the representation, the object that is represented and the social context within the representation that is being made sense of. Because these also shift over time, this triangle can be thought of as a Toblerone chocolate box, and because there are always more than one social representations of an object or a series of related objects, we should think of a stack of Toblerones.

Social representations can also facilitate the communication between groups, which may for example have anchored new concepts to different contextual knowledges or expectations; as I argued in Riesch (2010), they can act as boundary objects (in the classical science studies sense of Star and Griesemer, 1989) between two groups that understand a concept differently in detail but nevertheless manage to converse across the boundary through the common origins of the concept. In Star and Griesemer's phrase, the social representation as a boundary object can act as a "common coin" that facilitates intergroup communication and eases the cross-fertilisation of ideas. US evangelical biblical interpretations of Revelation, for example, managed to travel across otherwise jealously guarded social and communicative boundaries into contemporary Islamic apocalyptic literature (Filiu, 2011). Similarly, I will argue that millennial concepts of post-apocalyptic utopias or messianic salvation travel through the secularisation barrier and find a new home, and take up new meanings and connotations within a modern, areligious social context.

With this as a conceptual tool as a way of thinking about how representations evolve and influence our collective social epistemologies and communicative practices, we can try to understand how idea about apocalypse have travelled through the centuries and across community and social boundaries. Ideas formulated ages ago under different social contexts travel through time to acquire new connotations and adapt themselves to new circumstances; these in turn inflect how the old ideas are understood in retrospect.

One of the central arguments I plan to make is that apocalyptic narratives – whether about climate change, asteroid impacts or nuclear war – can be understood as Blumenbergian myths, or more narrowly, our current apocalyptic narratives are social representations that have been anchored on more traditional, often religious ones. In one sense, even older apocalypses are themselves social representations, for example the personification of disaster is neatly demonstrated through the traditional imagery of the four riders of the apocalypse – war, famine, pestilence and death – each of which is a (somewhat) abstract concept given reification through its representation as an identifiable person. This then also gives these abstract concepts a moral agency that lets us explain the uncaring, unfeeling action of things of chance as decisions and action made by someone.

Following social representation theory, the way we make sense of any new ways for the world to end will be rooted in our wider cultural narratives surrounding the end. This will vary from person to person, as there will be different anchor points in people's personal worldviews, though we might expect certain commonalities between groups from the same cultural backgrounds because it is the social context that frames our anchoring points and shapes our social expectations and knowledges. Variations will be even stronger between different social and cultural backgrounds.

Social representation theory has become a mainstream approach within the more socially oriented, qualitative subdiscipline of social psychology but has not found its (in my view) deserved, wider recognition within sociology. But the attraction of social representation theory to the study of the relation between science and religion is obvious, I think, as it provides us with a conceptual tool for how to see the ontological, epistemological and communicational interrelations and links between these two knowledge communities.

Chapter 5

What to know and how to know it

Introduction: prophecies and predictions

The history of apocalyptic prophecy is littered with disappointment (or relief) because the world has so far refused to end. Even if at times it seems to be having a very good go at it, civilisation has so far recovered and continued (though of course individual civilisations have indeed at times found their end). The fact that millennial movements nevertheless often hold on to their faith, despite having posited often precise and clearly defined end dates that passed without incident, is well established and the focus of one of the most foundational studies within both the sociology of millennialism and wider social psychology (Festinger, Riecken and Schachter, 1956). This stubborn refusal to accept obvious facts may be seen as one of the distinguishing features between religious knowledge communities and science (which, after all, relies on observing and recording clear evidence and adjusting its theories accordingly).

However, as scientific and religious knowledge communities are both (for the most part) earnest attempts at making sense of the world – and as I outlined in Chapter 2 a clear delineation between the two as knowledge practices is not possible – this distinction may not be entirely clear either. First, there may indeed be a deep rationality of the religious community holding fast to its belief in the face of clear evidence to the contrary, and second, science also does not generally work as rationally as it is popularly made out to be and often disregards clear evidence to the contrary of its core beliefs. And this is not so much a bug within science but one of the features of how science progresses, that is it can be argued that a certain amount of disregarding of evidence is actually a good thing for scientific progress.

As a way of looking at the connections between the apocalyptic narratives and contemporary "existential risk" apocalypses, this chapter will explore the epistemology of apocalypse. Specifically it will attempt to use (a particular) philosophy of science as an epistemological tool to investigate these connections. This is not to argue that apocalyptic prophecies are just as valid as scientific predictions (they clearly are not). Instead, I want to show that the social psychological foundations of how we in groups make sense of and interpret evidence are similar in different areas of human knowledge production and evidence of a wider folk epistemology from which science derives. Clearly, for example, both prophets and scientists make predictions, and people believe that observing these predictions come true will confirm or corroborate the prophet's or scientific theory's wider system. This chapter will first look at Festinger's very influential attempt at theorising how prophecy deals with failure and then link this to a larger attempt at a theory of knowledge based on Imre Lakatos' (1978) philosophy of science.

Festinger and the apocalyptic

Festinger, Riecken and Schachter (1956) start their influential ethnographic study on an apocalyptic UFO cult with the following observation:

> Suppose an individual believes something with his whole heart; suppose further that he has a commitment to his belief, that he is presented with evidence, unequivocal and undeniable evidence, that his belief is wrong: what will happen? The individual will frequently emerge, not only unshaken, but even more convinced of the truth of his beliefs than ever before. Indeed, he may even show a new fervor about convincing and converting other people to his view.
>
> (Festinger, Riecken and Schachter, 1956, p. 3)

Festinger has developed his theory of cognitive dissonance as a way of trying to explain this phenomenon. Cognitions, or beliefs we hold, can either be consistent or inconsistent with each other. When we hold two inconsistent beliefs, we experience "cognitive dissonance", a state which Festinger argues is inherently uncomfortable and something we aim to resolve: "There will arise pressures to reduce or eliminate the dissonance" (Festinger, 1962). We can resolve the dissonance by either abandoning a belief, tweaking your beliefs by adding reconciling beliefs or engaging in compensating behaviour. While it is of course impossible to clearly quantify this, Festinger posits that dissonance will be pronounced when the person experiencing dissonance has made clear sacrifices on the basis of their beliefs. Similarly, a person is less likely to change a belief to which there are *psychologically* committed.

Festinger, Riecken and Schachter (1956) observed the phenomenon in the opening quote above in the well-known historical case of the Millerite prophecy. William Miller, a Massachusetts lay Baptist preacher, calculated the end of the world to happen by March 1844, going by a calculation based on a verse in Daniel, and built up a large following of people, many of whom psychologically and physically invested a lot in this prophecy. When the date came and went, rather than seeing this as a disconfirmation of Miller's prophecy, further introspection determined that the calculation was off by a few months, and a new date was announced for October 1844. The passage of the second date without incident became known as "the great disappointment", and this did occasion a large number of followers to abandon Miller's prophecy. A sizeable number however, managed to carry on, through for example positing that the events as described in Revelation did indeed start in 1844, but that it will still be a while before they will become apparent. This then gave birth to new denominations (the various branches of Seventh Day Adventism), which now generally very wisely refrained from providing any more precise dates as to when the end will visible start. The failure of the Millerite prophecy then did not lead to uniform abandonment of the belief. Instead, while some people who were more on the periphery and less invested in the movement did abandon it, others felt their faith strengthened and continued it through the establishment of new churches and increased proselytisation.

Festinger et al. provided a list of five conditions (pp. 31–32) that need to be fulfilled for a failed prophecy to lead to increased fervour and proselytising: (1) the belief needs to be held with conviction and have relevance to the believer's behaviour; (2) the believer must have committed themselves by taking action that is difficult to reverse (e.g. in the Miller example, farmers who abandoned their crops because the end was near); (3) the belief must be sufficiently specific that events can irrevocably refute it; (4) such undeniable evidence must happen; and (5) the believer must have sufficient social support.

Intending to study the phenomenon of believers not abandoning their apoca-lyptic faith despite clear evidence to the contrary, Festinger and his co-authors responded to an advert from a small millennial UFO cult in a newspaper: Members of the team approached the cult and joined as members, proceeding to record the group's reactions to the failure of their predictions in their role as covert par-ticipant observers. The cult had predicted a cataclysmic flood to happen to the US and benevolent aliens with whom they were in contact would appear to rescue the cult members. When they failed to show up at the predicted time, most of Festinger's conditions were fulfilled: The members of the cult had a firm belief in the rescue, for which they made many sacrifices in terms of giving up jobs and alienating friends and family members, and the evidence (that the aliens didn't show up) was undeniable and recognised by the cult members.

The initial reaction was to add a reconciling belief as a way to reduce the disso-nance between the belief in alien rescue and the clear evidence of no such rescue: The additional belief in this case was that the aliens staged a dress rehearsal to ensure the cult members are ready to be picked up at a revised date. In the case of the Millerite prophecy, the failure of the apocalypse happening at the specified date was explained as being an initial error in the calculation of the date – in both cases, an additional belief satisfied both the overarching apocalyptic belief system and the disconfirming evidence.

When the rescue failed to turn up at the revised date as well, the reaction went two ways. The more peripheral members of the community, people who had made fewer sacrifices and who held to the belief with slightly lesser conviction, withdrew and essentially dealt with the dissonance by abandoning the belief. The more centrally involved members by contrast responded with increased fervour in their belief: It was now felt that their actions and steadfast belief prompted the indefinite postponement of the prophesised apocalypse. This reconciliation led to a renewed fervour by the cult members, and a group previously reluctant to attract publicity started actively promoting to the press the good news about the abandonment of the apocalypse.

Festinger et al. believed that this observation was a very good fit and thus a confirmation of their hypothesis. Although it has been foundational as a study of disappointed prophecy within millennial studies, there are also several issues with it that need to be noted. First, the empirical validity of the study design can be (and has been) questioned (van Fossen, 1988; Bainbridge, 1997). The cult under investigation was rather small in numbers, and with the relatively large group of social scientists having joined in as observers, it is likely that they will have had an unintentional effect on the group's behaviour (it also raises ethical issues around their use of covert observation). Also, a different take on the study can also be undertaken through by and large accepting the results but instead reinterpreting the findings through other social theories (Jenkins, 2013), indicating that we might not really need cognitive dissonance to explain what Festinger et al. observed (as with the UFO belief itself, Festinger's explanation of what happened is also underdetermined by the evidence).

Probably more damaging was the fact that several other studies failed to replicate the behaviour predicted by Festinger's theory, in particular the effect of increased proselytisation following the failed prophecy was not observed regularly in other similar circumstances. In Stone's (2000) collection, most of the case studies have differed from the findings of Festinger and his colleagues (e.g. Schmalz's 1994 study on Jehovah's Witnesses who appeared to retreat rather than proselytise, initially at least, after their failure to predict the end of the world in 1914). While this was a failure to replicate Festinger's expectation of increased proselytisation, other aspects of the Festinger scheme were however still present, notably the tendency of the movement members to find alternative explanations for the failed prophecy that otherwise leaves the main aspects of the faith intact.

Other criticism attaches to the whole concept of cognitive dissonance that Festinger aimed to establish experimentally (i.e. through research methods other than ethnographic study), with recent research on some of the classical experimental studies in social psychology having raised general doubts as to their reliability, potentially adding cognitive dissonance to the wider pile of the contemporary replication crisis in experimental social psychology (Earp and Trafimow, 2015).

While all this is indeed concerning, it does seem that the larger pattern of responding to failed prophecy through the alteration of auxiliary hypotheses seems to be correct, regardless of Festinger's attempted explanations surrounding cognitive dissonance, or some of his more precise expectations such as increased proselytisation, and the precise conditions under which this would happen. The fact that neither Miller, nor early Jehovah's Witnesses, nor Festinger's UFO cult accepted the failure of their prophecy and indeed found other reasonable explanations is undisputable and a matter of historical record, and this is observable in almost all other examples of prophetic failure. In this sense, though the details and precise cognitive processes involved are in dispute, the larger picture of the pattern is not.

Although the theory of cognitive dissonance is meant to be a wider explanatory theory of social behaviour and cognition, given the subject matter of Festinger et al.'s study of an apocalyptic cult it is unsurprising that cognitive dissonance theory has become one of the major cornerstones of social scientists studying the apocalyptic (see McGhee, 2005), being ranked alongside Norman Cohn's (1957) *The Pursuit of the Millennium* as one of the foundational texts of millennial studies by McGhee (2005).

In particular, cognitive dissonance was used as one of the main factors in Landes' "apocalyptic wave" account of the dynamics of how millennial movements develop (Landes, 2011) that I outlined in Chapter 3. Prophecy can fail in other ways than an unfulfilled apocalyptic prediction of the end of the world. A millennial movement might be promising the millennial reign of peace after a cataclysmic convulsion. Marxism, for example, promises an increasing worsening of social conditions until the time of the cataclysmic takeover of the proletariat, after which a utopian, millenarian reign of peace and justice will start. A communist revolution then will be faced with failed prophecy when the post-revolutionary world falls short of millennial peace. Similarly, the humanist tradition that led to the French Revolution promised a utopian vision of peace and justice with the overthrow of the ancien régime and was frustrated in that expectation too.

In line with the relative deprivation thesis, it appears to be a rule that a millennial movement needs to feel to be part of a minority, but also the whole apocalyptic theology often only makes sense if there is a wider, cosmic struggle between good and an overbearing evil. If an apocalyptic religion establishes itself, it will over time lose the apocalyptic element in its theology. St Augustine's amillennialism found a parallel of sorts in the loss of millennialism in Chinese communism under Deng Xiaoping or Soviet communism under Khrushchev. However, in the immediate moment of the revolutionaries finding themselves in power, almost unexpectedly so, with the erstwhile evil having been conquered, the dissonance needs to be resolved through the postulation that the fight has not actually yet been won and the forces of evil, be they reactionary capitalism, counter-revolutionaries, Jewish deep-state conspiracists or papist sympathisers, plot just behind the curtains to thwart the establishment of the millennial kingdom.

Thus when a movement bent on revolution and the fervent belief that things will be much better once the old and corrupt regime has been swept away has achieved its aim of gaining power, it is often found that the promised utopia is still in many ways not present. There is thus a dissonance between the fervently held expectations of a better world and the harsh and undeniable reality of life, especially in the chaos following revolution. This gives us an explanation of some of the most destructive and violent episodes of successful revolutions: The cognitive dissonance is resolved by adding reconciling beliefs, with the most obvious-looking one being that it is the old guard that actively works against the construction of the new utopia. The violence following the French Revolution that sought to eliminate sympathisers of the old rule or forcefully suppress any expression of such sympathy – the "Jacobin terror" – has parallels in the Russian, Chinese and Cambodian communist revolutions, Nazi Germany, the tyranny of John of Leiden in Münster, and most other successful cataclysmic overthrows of established social order.

The view from the philosophy of science

While on the face of it the differences between apocalyptic UFO cults and scientific research are stark, the motivating thought for Festinger, Riecken and Schachter, I believe, does have a clear equivalent in the study of science. As historical studies have shown and argued, most famously those of Thomas Kuhn (1962), it is rarely the case that a clear experimental observation refuting a scientific theory or hypothesis succeeds in dissuading scientists who have invested in that theory to abandon it.

My argument will depart from the observation of the similarity between Festinger and Kuhn (which by itself is not an entirely new observation; see McDonagh, 1976), and proceed by using this as a starting point to a wider theorisation on how insights from the philosophy of science might be useful to the study of the apocalyptic by looking beyond Kuhn to the philosophy of Imre Lakatos and his concept of the *methodology of scientific research programmes* (Lakatos, 1978).

Disconfirming evidence, even if usually less spectacular than the failure of the apocalypse to appear, also happens in science. How scientists react, or should react, to disconfirming evidence is one of the central questions in both philosophical and sociological studies of science. A somewhat simplified version of the major debate in the philosophy of science in the last half century might posit a conceptual move away from overly normative philosophies that view science as a rational enterprise following strict rules of logic, to the realisation that historically, science and scientists simply haven't acted like this. Popper's *Logic of Scientific Discovery* (2005 [1959]), arguing for an orderly process of hypothesis and theory construction leading to clear predictions about future events that can then be tested and, if falsified (i.e. disconfirmed), rejected, has won many converts, often among scientists themselves. This was influential to the point that Popper has become symbolic of how scientists view their activity, even if in practice their activities and thoughts diverge quite some way from Popper (Riesch, 2008).

On a logical level, however, this has been shown not to be quite as clear as that. As W.V.O. Quine (1980 [1953]) and before him Pierre Duhem (see Gillies, 1998) outlined, a theory or hypothesis never exists in isolation and relies on a range of statements and background assumptions, most of which can be changed to leave the theory essentially unchanged when disconfirming evidence is encountered. Philosophers refer to the Duhem-Quine thesis as the *underdetermination thesis*: No theory is ever completely determined by the evidence; in other words, given the evidence that we have, there are always a potentially infinite number of theories that are consistent with our observations.

On a historical level, Thomas Kuhn (1962) has famously shown that scientists have often overlooked disconfirming evidence in order to keep with their theoretical framework or "paradigm", at least until the point comes that the disconfirming evidence has piled up so as to become increasingly uncomfortable, when eventually a revolutionary shift to a new paradigm might occur.

A lot of the subsequent history of this debate within the philosophy of science revolves around either trying to find ways in which the underdetermination thesis as well as scientists' actual historical behaviour can be understood as a rational and methodological enterprise after all. Within the field of the sociology of scientific knowledge, which was inspired in part by Kuhn's, the hope of finding normative rationalisations of science has been given up in favour of historical and empirical investigations of how scientists react to disconfirming evidence, rather than how they should react. In both fields, though, investigations have now moved on to other grounds, but the question of how to theorise disconfirming evidence has not been settled yet in either field.

Within the brief post–Kuhn/Popper philosophical tradition that aimed at reconciling Popper's rationalisation of scientific logic with Kuhn and Quine's logical and historical criticism, the work of Imre Lakatos stands out as particularly influential. As a philosopher, Lakatos was rather dismissive of sociologists of science (Gillies, 2014). However, what he has provided can in some ways be seen as a clearly sociological theory – albeit normatively sociological – about how knowledge can progress in an orderly and rational procession while still acknowledging Kuhn's insights.

My argument will be that we can treat Lakatos' approach as a general sociological or social psychological structure that applies to the structures of human knowledge claims generally rather than singling out science as a special case, and that it will help provide explanatory force to the development and dynamics of the epistemic communities concerning themselves with apocalyptic prophecy.

The point of this exercise is twofold. First, I want to use this as a sociological comment on the difficulties of separating scientific modes of knowledge and belief construction from other socially constructed ways of knowing. Second, I believe this can make a contribution to the sociology of millennialism by using insights from the philosophy of science and applying them to a new field within the sociology of religion: On a larger level, scientific and religious ways of knowing are not as opposed epistemologically as often assumed. This will, I hope, have the added bonus of adding a re-examination of Festinger's study and its relevance to millennialism, because it takes the epistemologically relevant parts of Festinger's theory and strengthens it through Lakatos' specific philosophical contribution.

Imre Lakatos (1978) developed a philosophical account of science as a direct response to the Duhem-Quine thesis and Kuhn's demonstration that historically successful science has developed rather differently than the clean logical and dispassionate model of "falsificationism" that Popper proposed. Lakatos' initial name of "sophisticated falsificationism" shows the clear intellectual heritage as an intended update of falsificationism in the light of Kuhn and Quine's challenge. This did not go down well with Popper, as it also implied his system was a sort of naïve falsificationism, and though the two philosophers had irrevocably fallen out over this (and other issues), Lakatos' system has since become known rather through the less divisive and more descriptive name of the *methodology of scientific research programmes*, mostly published posthumously after Lakatos' untimely death in 1974. Although enormously influential within the philosophy of science, Lakatos never acquired the same recognition among scientists and sociologists of science as did Popper and Kuhn.

The challenge that Kuhn and Quine threw at the empirical philosophical tradition of Popper (and the logical positivists) was that disconfirming evidence could relatively easily be accommodated within a wider theoretical framework without the need of scientists to abandon it. This is both a logical consequence of how scientific theories are structured (with lots of background hypotheses and assumptions that can be tweaked in response to disconfirmation) and a historical fact. Lakatos tried to propose a logical structure to science that accommodates these features but still demonstrates a clear methodological progression that can explain why science still manages to work as successfully as it has done so far.

The main units of analysis for Lakatos are not individual scientists or individual scientific theories but instead "research programmes" composed of scientists who work within an interpretative scientific framework. Comparable in ways to Kuhn's notion of a paradigm, this framework consists of a set of hypotheses and assumptions that together make up the theory that the scientific research programme works with. Similar to the idea of the paradigm with Kuhn, within the research programme, scientific methodology is applied and evidence interpreted in accordance with this wider guiding framework. The hypotheses and assumptions of the framework come in different types of importance to the whole programme: A "hard core" of hypotheses that defines the general interpretative outlook of the programme which cannot be changed or abandoned without the programme losing its essential character, and a "protective belt" of additional assumptions and hypotheses which can be changed in response to disconfirming evidence. Importantly, as Musgrave and Pigden (2016) outline,

> these modifications are not random – they are in the best cases guided by the heuristic principles implicit in the "hard core" of the programme. A programme progresses theoretically if the new theory solves the anomaly faced by the old and is independently testable, making new predictions. A programme progresses empirically if at least one of these new predictions is confirmed.

An historical example given by Lakatos as a reinterpretation of Kuhn's own history of the scientific revolution (Lakatos and Zahar, 1976) was Copernicus' heliocentric system: Though successive empirical observations have brought massive changes to Copernicus' system, the hard core idea that the sun is in the centre was unshakable, and subsequent observations were accommodated into that basic assumption by making tweaks in the protective belt of hypotheses (such as that planets moved in circles, which was abandoned after Kepler described the planets as having elliptical orbits).

Thus a research programme will never as such suffer a fatal blow delivered by evidence, as the protective belt can be adjusted in order to save the core. However, the more often this needs doing, the worse it looks for the programme: Lakatos introduces the concept of a degenerating research programme which struggles with streams of disconfirming evidence coming in and making it an increasingly laborious task to change background hypotheses. While a research programme can easily cope with changing protective belt assumptions occasionally, it is a bad look if this becomes all it is doing. A progressive research programme by contrast keeps receiving evidence that accords with both core and protective assumptions, even if it may occasionally need to tweak some protective assumptions. Research programmes in the same scientific area can coexist and several can thrive at the same time, however eventually the dynamics will be that one research programme stays progressive while the others degenerate. A scientific revolution happens if a previously dominant and progressive research programme degenerates to the point that it gets abandoned completely and scientists change over to newer, progressive programmes, which in turn may stay progressive for a while but also degenerate over time to be superseded by the next generation of progressive research programmes, and so on. Research programmes can exist in competition with each other, and even if one research programme has achieved overall dominance, scientists can still work within the framework of a degenerated research programme trying to make it progressive again.

As Musgrave and Pigden (2016) point out, the fact that the modifications in the protective belt don't happen randomly is a key point in Lakatos' project. As they put it:

> Even if it is sometimes rational to persist with the hard core of a theory when the hard core plus some set of auxiliary hypotheses has been refuted, there must surely be some circumstances in which is it [*sic*] rational to give it up! The Methodology of Scientific Research Programme has got to be something more than a defence of scientific pig-headedness!

A progressive research programme must be both theoretically and empirically progressive. Theoretically progressive means that the programme must be able to predict "novel and hitherto unexpected" facts, and it is empirically progressive if (at least some) of these facts are corroborated by observation. It is degenerating if it is not either theoretically or empirically progressive, that is if the programme does not manage to predict any more novel facts, or if the novel facts it does predict keep getting falsified by the evidence.

This model retains an essentially logical view of the methodology of science incorporating Quine's insight that theories are underdetermined by the evidence, in that within a research programme new evidence gets incorporated according to the main objectives of the research programme while at the same time accounting for progress of science over time. Kuhn's concepts of paradigms and scientific revolutions also receive an explanation as well as a more considered account than Kuhn gave of how paradigm changes lead to progress in science rather than just the continued overthrow of mutually incompatible ways of understanding the world.

Lakatos' scheme is a normative philosophical account of how science should run, but by trying to explain the historical success of science so far, it is, like Kuhn, also an account of how scientists behave without consciously following Lakatos' method. Instead, per Lakatos, this epistemology is essentially social and emergent, something that scientists have traditionally done without being aware of it. The question thus arises of how much this is behaviour that is limited to science, or how much this might describe a general social psychological folk epistemology that drives general human attempts at understanding the world around them.

Lakatos and apocalyptic prophecy

In order to build a philosophy of science inspired theory of millennial and apocalyptic knowledge we can start by re-imagining the core unit of analysis – scientific research programmes – as the interpretative repertoires, beliefs and knowledge claims as well as the social norms and values that make up a social group's collective identity. This is not as particularly great a leap and echoes, for example, McDonagh's (1976) identification of Kuhn's paradigms, Lakatos' research programmes and the collectives of interconnected cognition that social psychologists such as Festinger (and others) were analysing. We may call this collective of a group's beliefs, knowledges, norms and interpretative frameworks that form part of the group's social representation of reality (loosely after Moscovici, 1981) an epistemic group.

The social representation of a group will be made up of a core of beliefs that define the group's collective identity (Tajfel, 1974): These are the tenets that define the group's sense of itself and as such includes the foundational beliefs that are most crucial to the group's self-identity and which cannot be abandoned without the group losing the essential character that defines it. Lakatos' example of the Copernican heliocentric system could here be joined by the UFO cult whose central, identity-defining "hard core" included the existence of the benevolent aliens and the fact that they had been communicating with the group as well as the impending doom and rescue, whereas the protective belt included the detail of specific dates. More destructively, the hard core of the revolutionary Jacobin social representations of reality included progressive enlightenment ideals that promised a better world, where their protective belt to be modified included their modification that they had maybe not been completely successful yet in their takeover of the country and that therefore extra effort had to be expended to root out the aristocratic holdouts.

So far, merely relabelling theoretical concepts from a Festinger-inspired social psychology of millennial movements in terms of Lakatos' philosophy of science is not particularly insightful other than as another potential example that the social and philosophical sciences have a habit of reinventing the wheel without realising, because the subject matter the two areas study (science and millennial cults) are rarely analysed within the same social scientific subdiscipline. However, we can try to discern features of one system that do not have clear analogues in the other and see whether this brings any new insights. On viewing Lakatos as a specific formulation of the social psychology of science there have already been a few, though by now somewhat forgotten, contributions (e.g. McDonagh, 1976).

Going the opposite way, however, seeing the parallels between a system specifically designed to rediscover normative scientific rationality within the messy and often disordered tussle of real-life science and the social identity of seemingly delusional and often socially destructive religious and quasi-religious cults and revolutionary movements might bring some interesting insights: If general epistemic groups, and particularly the spectacularly refuted social movements of apocalyptic prophecy cults, behave in ways similar to science, what are the implications to our hopes of finding a special place for science as a uniquely rational activity? But what might be a loss for Lakatos' hopes of finding a special place for science might become a gain for students of the apocalypse and its associated movements.

In a millennial cult, the predictions (prophecies) will similarly be novel and express hitherto unexpected facts – first, if the prophecies were only to express predictions that we expect anyway, the cult would not be successful in distinguishing itself from mainstream societal expectations. Second, as with scientific research programmes, the most spectacular and eye-catching prophecies will be those that are specific enough to be tested empirically – the world will end on a specific date rather than at some undetermined (and therefore untestable) time in the future or even in a metaphorical time outside of real time, as argued by theological moderates like St Augustine. Cults that manage to engage the imagination of new followers are therefore invariably theoretically progressive, almost by definition of what is commonly understood as prophecy.

An empirically progressive prophetic cult by analogy would need at least some of the prophecies to be corroborated. While of course so far, no prophetic cult has yet correctly predicted the end of the world, the corroboration of at least some of its other predictions would qualify it in Lakatos' terms as empirically progressive. And this is indeed easily and often observed: As historians of millennial cults such as Cohn have demonstrated, almost any period in history has a profusion of prophets with varying messages and predictions on smaller matters as well as the eye-catching, end-of-the-world ones. Some of them will sometimes make clear predictions on other matters that happen to turn out corroborated, through chance because of the sheer number of prophets, but also because some of these predictions are not quite as novel and unexpected as they would appear in isolation. Even within one particular movement, prophecies can multiply enormously. Robertson (2016) introduced his concept of "rolling prophecy" in the context of modern conspiracy theory, where with the profusion of prophecies made, invariably one or two will come true and therefore act as corroborating evidence, while the rest are then easily forgotten or downplayed. However, the corroboration of at least some prophecy lends credence to the whole enterprise.

Lakatos clearly means empirical progressiveness to be the corroboration of the novel predictions in the research programme rather than its run-of-the-mill ones. This has a clear parallel in prophetic predictions, where also the corroboration of the more unexpected predictions will more successfully convince followers and prospective followers of the truth in the whole programme. Skilled (or lucky) prophets will therefore be able to convince their followers that fairly ordinary predictions are in fact surprising.

All this is not to say that a good scientific research programme is successful because it makes many novel predictions and hopes to get lucky in at least some of them, as in the cynical interpretation of Robertson's (2016) rolling prophecy concept. However, the persuasiveness of rolling prophecy relies on the same mechanisms as the persuasiveness of corroborated predictions in scientific research programmes. And this is a subconscious effect – people who engage and believe in rolling predictions are not necessarily cynical themselves but actually feel this shows their theories to be corroborated.

Thus it appears that the semiotic excitement that Landes (2011) observes, where the prophetic movement sees corroboration of their belief system almost anywhere they look, emulates empirical progressiveness as far as the followers are concerned. Whether prophetic cults are ever really empirically progressive in Lakatos' sense of a successful scientific research programme or just got lucky and appear so in the eyes of semiotically excited followers may probably be one crucial distinguishing feature of a cult from science. But significantly, it won't be readily recognisable as such by the people invested in either the cult's or the scientific programme's social identity.

In any case, just as Lakatos tried to describe with science, at any one point we will have several competing prophecies and prophetic groups; usually one will be dominant, but at times there may be several competing ones vying for dominance that might seem equally plausible to prospective followers. Dominance will be decided by their progressiveness in terms of how much they are theoretically and (seemingly) empirically progressive. We can therefore start to build a model of the dynamics of competing prophetic movements that complements the internal dynamics of the waxing and waning of apocalyptic movements that are described by Landes. Landes' model is – with the exception of his use of cognitive dissonance theory – rather descriptive and reliant on personality traits (in his terms, people involved are either owls or roosters) that by themselves do not get much further explained. The apocalyptic wave "breaks" when the failures of the predictions start to pile up and eventually reasonable voices (owls) start dominating the discourse again.

Concluding thoughts: beyond prophecy versus prediction

Scientists make predictions and prophets make prophecies; both activities are projecting expectations into the future based on wider epistemic knowledge frameworks and belief systems that have been build up through previous observations and traditions from the past. In theory at least, the predictions and prophecies will be compared with the evidence as it plays out, and the scientists' and the prophets' epistemic framework will be evaluated positively or negatively, depending on how well the evidence ends up matching with the expectation.

The exact manner of how this works has been a matter of contention and debate. In the philosophy of science Lakatos' scheme is well respected but certainly not seen as the ultimate account of science – this, if that means an account on which a majority of philosophers agree, does simply not exist. Similarly, what exactly prophecy is and how it works is an ongoing matter of debate within philosophy and sociology of religion. My observation here, which I don't think should be controversial, is that they are both concerned with projecting expectations into the future, with the expectation that their occurrence or lack of occurrence will have an impact on how we should evaluate the belief system that gave rise to these expectations. And of course, there will be differences in the details – I don't intend to claim that prophecy and scientific prediction work exactly the same way, merely that they are both types of the same sort of thing, an instinctual social epistemology by which scientists and believers – all of us basically – try to make sense of the world.

Evaluating how these events play out socially from outside of such belief systems shows up differences in the reliability of the knowledge that is being produced (and from that perspective science tends to look better than, say, UFO prophecies); however, crucially it looks very different from the inside. A prophecy makes predictions for the future, people believe in it as fervently as their initial belief to the overall belief system warrants, and that reaction is then perfectly rational on its own. Evidence that disconfirms the prophecy can then, again perfectly rationally, be explained through altering less important facets of the belief system, leaving the overall system intact.

Chapter 6

Risk

The many meanings of risk: risk as a boundary object and social representation

As noted in Chapter 4, current academic and scientific debates around potentially world ending events are often phrased in the language of risk, with "existential risk" being the research object of recently established centres such as the Future of Humanity Institute in Oxford or the Cambridge Centre for the Study of Existential Risk. As a rather different conception of risk within the discipline of sociology, Ulrich Beck's (1992) work on risk as a defining part of late modern society has become part of a wider and much utilised social theory. This conception of risk has, particularly through the later work of Beck (2007), found its application to the large global risks such as climate change that motivate the existential risk researchers.

The concept of risk, meaning different things within these two contexts, performs some wide-ranging rhetorical work here with respect to how we think about the future. This chapter will look at how risk is being conceptualised as a general concept within social science and link this to the current discursive practice that foregrounds risk as the main analytical lens through which we think about the future, noting that risk is a contemporary frame through which apocalyptic fears are being expressed. Since I am writing this chapter in the context of the 2020 coronavirus crisis, and because infectious disease appears to be particularly relevant for one of the theories of risk I will be examining, the pandemic (and the risk it represents) will be my running example. First, however, I want to provide an overview of the concept of risk and the various meanings and discursive functions it has been put to work on.

Risk is a very malleable term that has been used in various different senses in various academic disciplines and contexts within individual disciplines as well as within ordinary language. This interpretative flexibility of the term is clearly one of the reasons for its conceptual success. As with several other rather vague but important and ubiquitous terms discussed in previous chapters – science, religion, the apocalypse itself – this vagueness means that various actors can project their own slightly different meanings and interpretations onto it and transform it into the concept that is useful for them while at the same time projecting to the reader or listener the meanings and interpretations that they associate with it.

It is in this sense a boundary object, as influentially analysed by Star and Griesemer (1989) within Science and Technology Studies. A boundary object is a concept about which two (or more) different epistemic groups can have somewhat different interpretations and ideas but which they both recognise as important, which as such then allows them to converse across their epistemic differences; it is, in Star and Griesemer's phrase, a common coin. As I noted in Chapter 4, Moscovici's theory of social representations can be used to provide a wider interpretation of boundary objects; two or more epistemic groups will understand new concepts through anchoring them to their own background knowledges, assumptions and ways of seeing the world. As social representation theory outlines, this is a mechanism that explains how different groups can have radically different interpretations of a concept but also how they nevertheless can have meaningful conversations between themselves about it without ever agreeing what it really is.

In mostly scientific or technical literature (rather than the sociological or everyday conceptions of risk), risk is a measure of the probability of an event multiplied by the severity of its outcome. In this traditional understanding of risk that has more or less been lost in contemporary language, risk could therefore also apply to positive events, that is positive outcomes as well as negative ones (see Hacking, 2003, for a brief history that also links risk to the issues introduced here later).

More contemporary, wide-ranging usages of the term "risk" by contrast tend rarely to apply directly to calculating probabilities and outcomes of events and instead refer to a vaguer understanding that future adverse events may happen. This is not just because putting clear numbers on the probability (and severity of outcome) of vaguely defined future events such as climate change or artificial intelligence taking over is almost impossible; it is also because the meaning of the term itself in popular language appears to have shifted.

In this sense, a precise numerical definition is quite impossible but also makes no particular sense. Even if we agree on what the measure of that calculation should be in the first place (monetary value is clearly inadequate when human lives are at stake), and while we can put a monetary value on human life, this effort is complicated, subjective and open to much legitimate criticism. However while existential risk researchers such as Ord (2020) provide some reasonable attempts at quantifying various late modern existential risks, it does not strike me that these are really necessary for the force of their arguments; vaguer words such as likely, very likely or unlikely are more than enough to give us some indication of how worried we should be about them. In everyday life we can talk about risk and the risks of putting ourselves at risk without necessarily needing to know, or even strive for, having a clear number.

In the everyday meaning of risk, then, we talk about a vague foreshadowing of the future, which is imbued with all manner of different meanings and interpretations but which all in some sense fill us with dread. Beck's (1992) sociological concept of risk thus performs a different type of function within contemporary discourse, where the sociologist is maybe less interested in the actual outcomes and probabilities and instead looks to how societal awareness of these larger societal risks has shaped the development of public discourse on our future.

Thus while the concept of risk as first developed by and for the insurance and gambling industries involved concepts of probability and severity of outcome, the risks that are in the forefront of our worries in contemporary late modern society cannot be as easily brought into our conceptual calculations. The problem with risk as a measure of probability and severity of the outcome is in trouble under circumstances where the probabilities are difficult or impossible to assess, and/or when it is equally unclear what the severity of the outcome is or what we should even measure in order to assess it.

On the first issue, the problem with probabilities or uncertainties is that only in rare, slightly contrived situations can we give a fairly certain estimation of a probability, such as in a game of chance. The probability that we roll two sixes with fair dice can be calculated, and the problem of how we know whether the dice are fair can be investigated itself by rolling them a large amount of times. The uncertainty in more complicated events, where contributing parameters are not known precisely needs to be modelled, and then therefore depends on the certainty which we attach to the parameters we put into the model, and the certainty we attach to how closely our models match the underlying reality – as well as the certainty the consumer of the model attaches to the competence (or even honesty) of those doing the modelling. In complicated situations, then, our uncertainties are themselves subject to a large amount of uncertainty, making it difficult or impossible to attach clear probabilities to them (see Riesch, 2013; Spiegelhalter and Riesch, 2011).

On the second issue, the severity of the outcome, we face similar if not worse problems. The expected outcomes depend just as much on the parameters we set, the models we choose and the modellers we choose to trust, but added to that it is not always clear what kind of thing the outcome is, and when there are several types of outcomes, how they compare to each other. In games of chance, or even in insurance calculations, the outcome is monetary and thus fairly clearly defined; however in globalised late modern risks such as climate change or infectious disease outbreaks, we might prefer to think about the potential loss of lives as the main measure. However, monetary considerations are not unimportant even if the loss of lives is to be the main measure – for example, spending a lot of money on saving one life that could have been spent on building a hospital and/or training doctors and nurses so that potentially many more lives could be saved, would make little sense. But how then do we balance these costs? While sophisticated algorithms exist that put a potential price on a life exist to ensure overall maximisation of lives saved, where these prices are set exactly is ultimately always an arbitrary policy decision, since the uncertainties involved are ultimately too uncertain themselves.

Through plotting a graph of our certainty over the probability versus our certainty over the severity of the outcome, Stirling (2007) has attempted to put some order into the various risks we face as a society. When we are (fairly) certain about the probabilities as well as the outcomes, we face "traditional" risks that can be dealt with through statistical risk analysis. Scenarios where our knowledge of the probabilities is certain but not the outcome he terms "ambiguities"; the converse, where outcomes are known but not the probabilities, he terms "uncertainties". The most difficult risks then are what he calls conditions of "ignorance", that is where neither the probabilities nor the potential outcomes of an event are known. Mapping risk as a measure of uncertainty and (severity of) impact, Funtowicz and Ravetz (1990, 1993) distinguished different types of associated science policy scenarios: Where uncertainty as well as impact are low, policy works as normal science (a term they adapted from Kuhn, 1962), but when either uncertainty or impact are very high, we are in a "post-normal" policy scenario.

It is these types of risk, as well as Stirling's "ignorance" and the more intractable "ambiguities" and "uncertainties", that characterise the late modern risks that Beck is worried about and that will be relevant to my arguments on societal reactions to the potentially world-ending risks of the apocalyptic imagination.

In the following sections, I will introduce two influential sociological/cultural theories of risk, which will then be discussed with respect to the apocalyptic risks of interest to me. I will then present a third sociological perspective on risk inspired by the social identity theory; these three approaches will then be used to construct my own wider argument on why the sociology of risk is important in any contemporary discussions of the apocalyptic, as exemplified by the "existential risk" concept through which current apocalyptic fears get discussed.

Risk society

One of the most influential recent approaches to social theory has been Ulrich Beck's concept of the risk society (Beck, 1992, 2007), a similar concept to the theory developed by Anthony Giddens (first independently and later in collaboration; Giddens, 1991, 1999; Beck, Giddens and Lash, 1994), though I'll be concentrating on Beck here. Risk society theory posits that risk has become a defining feature of late modern (i.e. contemporary) society, and as such it has become a central feature in how we live and understand our lives and our relation to society.

In the transitional period as we move from modernity to late modernity,

> the production of wealth is accompanied by that of risks, which have proliferated as an outcome of modernisation. The central problem of western societies, therefore, is not the production and distribution of "goods" such as wealth and employment in conditions of scarcity . . . but the prevention or minimization of "bads"; that is, risks.
>
> (Lupton, 1999, p. 59)

Beck aims to show how risk as a concept has evolved over the centuries by dividing its development – or the development of our concept of future uncertainties – into

three distinct phases. Pre-modern (pre-industrial) society is characterised by seeing the future as incalculable and subject to fate or God's will. This then changed
in the industrialising societies through the development and refinement of concepts of probability, which placed future events in the realm of the knowable, pre-
dictable and thus controllable. Late modern or post-industrial society, however,
has started to understand risks that are non-calculable and thus barely controllable; they are not localised and can be global in nature and hard to assess.

The risks of late modernity that are central to Beck's thesis – nuclear disasters, extreme weather events, environmental degradation, biodiversity loss and
of course climate change – are all characterised by the basic unknowability of
precisely how likely these events are and what precisely will occur when they
happen. This is not of course to say that good and valuable research can't give
us an approximate idea of these events: We know that disastrous climate change
is a likely event and that if it happens, the outcome will indeed be disastrous.
Even the concept of the disastrous event itself has become diffuse and uncertain,
with late modern risks often becoming diffuse, "open-ended events, rather than
events that have a foreseeable end" (Lupton, 1999, p. 62). Thus we as a society
need to theorise them differently than risks that can be calculated more directly,
and through their nature they have a different social impact on how our society
structures itself, and they take on particular forms in how they are being discussed in public.

The nature of late modern risks is mostly scientific and/or technological in origin. This is to say that while science cannot give us precise information on them,
it is thanks to science and technology that we are aware of them in the first place,
and in some cases, it is science and/or technology we have to thank for them. This
is part of what distinguishes them from the risks that societies have always faced
throughout history. Of course, infectious disease outbreaks have always been a risk
to societies, however with (late) modern scientific tools, we have now become
aware of them in a much more intimate fashion; coupled with the transition of our
future imaginary from fate or God's will to an unknowable future characterised by
probability and chance, these risks, even though they have always existed, have
taken on a new type of meaning within our collective imaginary of the future.
Some of these risks, however, for example those of climate change or technologically mediated environmental degradation, are distinctly features of industrialism.
Due to their globalised, uncertain nature, late modern risks "are far more apocalyptic than in previous eras, threatening the destruction of all life on earth" (Lupton, 1999, p. 63), and this can be seen in their reappearance in the new discipline
of existential risk research.

Despite their scientific/technological nature, the incalculability of late modern
risks leads to ambiguities on how they should be responded to; characterised by
disagreement between experts and scientists and the additional uncertainty on
how policy should respond to them even if the experts presented a clear picture,
this then leads to the risk response moving out of the scientific and into the clearly
political arena and a paralysis of action.

The move away from fate or God's will in our imaginary of future events brings with it a change in who is to blame for disastrous events: The scientific/technological nature of late modern risks in particular shifts the blame or responsibility to humanity itself, and in some cases, to individual people (linking late modern risks to the Prometheus myth). Risks such as climate change or nuclear war can be seen directly as the responsibility of humanity, however even events that would previously have been attributed to fate, chance or "acts of God" are reinterpreted as *someone's* fault or responsibility. Thus, for example, infectious disease outbreaks – which have been a constant feature of civilisation throughout history – are reinterpreted as caused or at least exacerbated through human fault. Popular discourses on the coronavirus crisis, for example, heighten variously the xenophobic imagination on unhygienic practices in other countries as leading during the early stages of the outbreak or the failure of people to respect containment measures such as quarantine and social distancing in the later stages of the outbreak. (At the time I was writing the first draft of this paragraph, in March 2020, the major public discussions were focussed around China, where the virus originated, and thus were accompanied with popular, often clearly xenophobic, speculation about how China and Chinese people allowed it to happen. By the time of rewriting for the second draft, the focus appears to have shifted away from China, since the virus is now more clearly a global problem; however whole societies such as this time the US are still in the public spotlight for how their (mis) management of the crisis endangers the whole planet.)

This doesn't mean that blame responses are necessarily irrational or wrong: While the xenophobic elements of coronavirus discourse are clearly misguided, human responsibility can have an effect in how an event pans out. The adverse outcomes of extreme weather events, such as the high death toll in New Orleans after Hurricane Katrina in 2005, can likewise be linked to human responsibility (Beck, 2007) – not just through linking the increased frequency of extreme weather events to climate change but also the political failures that accompanied not keeping the levees in good enough shape or maintaining a health care system that would have been able to deal with the consequences better.

Nevertheless, despite the propensity for attributing blame in late modern risk events, their diffuse, unknowable and open-ended nature also makes it almost always impossible to clearly identify where the responsibility really sits. No single person, nor even one single clearly identified group of people, is responsible for climate change, nor is one identifiable person responsible for the nuclear bomb: If J. Robert Oppenheimer and his team hadn't built the bomb, it is probably reasonable to assume someone else would have done so (and the not-unreasonable fear at the time was of course that this someone else could have been German).

Most of the time, however, especially for the big all-encompassing global risks such as climate change, the blame lies not with individuals or individually clearly delineated groups but with global society as a whole. The self-awareness that global late modern society exhibits about its own collective responsibility for the hazards it causes to itself then constitutes one of the major hallmarks of (global) risk society as Beck sees it, that of its reflexivity. This means that society must and does confront itself as it responds to risk; it is aware of its own role in its potential destruction.

At the same time that risks have become global, however, Beck also argues that they force us to individualise our own risk biographies. Late modern society is at its basis an individualised society, with every member responsible for their own risks. This individualisation is the other side of globalised late modernity and a consequence of the shattering of the old certainties of industrial society. The structure of personal life has therefore also become an issue of risk, with us individually now being responsible for our life choices and life trajectories. Education, career choice, sexual orientation and so forth, all of which were more fixed in preindustrial and industrial societies, are now our responsibility, and if things turn out wrong, this is also our responsibility. At the same time, the choices we make often themselves are becoming less stable, work and careers are less secure, marriages are no longer for life, work life and family life make contrary demands on each other and with it emotional and financial insecurities become the norm.

The risk society in the time of the pandemic

While there are many more nuances to the risk society thesis than this brief overview can do justice, we can still try to trace how this understanding of contemporary society links to the apocalyptic scenarios we currently face. Beck has himself extensively applied his theory to the big globalised risks of climate change, but as I am writing this in the spring and summer of 2020, an obvious place to start with is the currently global apocalyptic sentiment of the coronavirus pandemic.

The first difference to note is that while crises like climate change or the possibility of nuclear war are clearly consequences of scientific and technological progress, infectious disease outbreaks are at first blush at least a much more natural occurrence, one that has been a constant threat throughout human history. On the contrary, it is scientific and technological progress that is aiding us considerably in our response to the crisis, making it much less deadly than it could otherwise have been.

As outlined above, however, as with the case of other natural disasters like hurricane Katrina, human fault is still the main narrative through which the virus is being discussed. Despite it being a mostly natural disaster, media reports as well as popular discourse is, at the time of writing, awash with narratives of blame. As with other infectious disease outbreaks, the country of origin, China, has been singled out as being largely responsible, at least in the early stages. Even before it had been established how exactly the virus became first transmitted between humans and animals, xenophobic discourse on Chinese eating habits or hygiene practices holds that culture to blame, and the authoritarian governments of Hubei province and China as a whole have been criticised for suppressing knowledge and dissent in the early stages of the outbreak.

Blame and responsibility however also lie with the individual – from the panic buying and exhortations on hand hygiene of the early stages of the outbreak, to the social distancing and isolation requirements that start to appear as the new social norms as the situation started to be settling into the long term, to finally the current anti-mask and anti-vaccination countermovements that are becoming the new folk devil of the wider blame narrative.

Reacting to a relatively unknown and novel disease under great time and resource pressures, medical scientists and epidemiologists grapple with large amounts of uncertainty and are unable, at this stage, to provide clear enough guidelines that can be translated into decisive policy action. Policy and expert differences in opinion, within individual countries but also between them (when the global nature of the situation would ideally require an internationally coordinated response), exacerbate the helplessness we feel in the face of the outbreak. This is not to say that there are no clearly wrong responses on the policy level, where in some countries and communities clearly misguided responses such as ignoring or underestimating the problem appears to have impeded efforts at responding adequately. For the most part, however, even within the expert community the uncertainties at this stage are large enough to make a final assessment of who is right only really possible with hindsight. As Beck has highlighted, the complexity and intractability of late modern risks (or at least, of natural risks but understood through a late modern reflexive scientific perspective), moves the societal risk response away from the experts and into the political realm. We are, as Stirling (2007) would maybe characterise it, operating under conditions of ignorance, with high levels of uncertainty over both the probabilities involved as well as what the potential outcomes are – all, as Beck noted, in a situation where it's not even entirely possible to say what the precise nature of the event is that we are so uncertain about.

In essence, then, while the nature of the event we are facing at the moment is one where an ostensibly natural biological disaster that can happen (and has frequently happened before) in modern and pre-modern societies, our social interpretation of it is uniquely late modern in appearance. Our awareness of the inherent scientific uncertainties is greater, the policy uncertainties are consequently more visible, the globalised nature of the event makes a concerted policy response unavailable or at least unlikely and in the secularised absence of fate, responsibilities are to be found in the behaviour of authoritarian regimes (both in China and closer to home), our perceptions of their for us alien cultural practices of animal husbandry and food culture mediated through xenophobia, selfishness of our compatriots who stock up on toilet paper, or disrespect the newly developing social norms of social distancing, mask wearing and staying at home. Blame gets heaped on the political response, for not ordering lockdowns soon enough (or ordering them too soon), for not compensating people who lose their livelihoods through respecting the quarantine (or through endangering the economy by throwing silly money at them). Scientists have so far largely avoided any blame, but we might be seeing this later on if early hopes for vaccines turn out to be too optimistic or if we find out that with hindsight, the expert advice from epidemiology could have been different.

The main tenet of risk society as a social theory is that there have been large qualitative shifts between how late modern and previous societies perceive themselves, their (and their individual constituents') place in the world, and how they react and become aware of worldwide threats. Everything is different now, and our new global society has a new and unique perspective on itself and the world that has never been there previously. While I believe that there are many valuable insights in their characterisation of late modern society, it does however not seem to me as if the differences are as pronounced at all. Xenophobia in the face of new threats, blame and even notions of uncertainty and the unknowability of the future have been around for much longer, and deterministic ideas of fate are maybe not quite so different from the risks that late modernity is worried about. Other social theory on risk and its significance to society has highlighted the human universal at play through the consideration of other societies and modes of thinking.

Cultural theory of risk

The cultural theory of risk was a development of Mary Douglas' earlier work within social anthropology (1966, 1992) on risk and dirt taboos, which was then applied, in collaboration with the sociologist Aaron Wildavsky (1983), to the risk discourses of then-contemporary environmental groups.

Risk is introduced to a society through the violation of dirt and purity taboos. Dirt, as Douglas introduces the term, is "essentially disorder"; it "offends against order" (p. 2); violations of taboos around dirt therefore introduce disorder into a society. Dirt and purity taboos tend to be in origin about lowering risks to a society through prohibiting uncleanliness, risky behaviour or certain foods that are perceived as risky and that threaten to destabilise them (Lupton, 1999, p. 39). Thus, for example, rotten food is clearly dirty, as it can harm people who eat it. Pointing towards medical origins of purity laws has a long tradition; Douglas for example points out Maimonides' attempts at finding rational explanations behind the food laws of Leviticus.

But purely medical explanations for purity laws, Douglas argues, are not sufficient in explaining what the laws do, where they come from and what function the play within a society, and even Maimonides struggled to find and explanation for why pork should be considered dangerous. Many purity laws throughout the various societies and cultures that Douglas looked at suggest that more is going on. Instead, risk avoidance strategies become ritualised. Whether any such things do cause potential harm is often not particularly clear or important to our reaction to the taboo, so that practices can continue being taboo-breaking even if they are not, or no longer, introducing a tangible risk. And since dirt is disorder, and an action that brings disorder is also dirty, disorder happens when things are not at their right place. Food is perfectly fine on the plate but becomes dirt when splattered on clothes.

Dirt and purity taboos can become highly ritualised; they would usually be based on tangibly dangerous behaviour, but it is the breaking of the taboo, the crossing of boundaries, that gets interpreted as the dangerous thing rather than the introduced danger itself. One of Douglas' own examples is the dirt taboos associated with shoes. Shoes by themselves and when used for their intended purposes are not dirty, but they can become so when put out of place: A shoe on the dinner table breaks the dirt-taboo boundaries of introducing the danger of whichever germs the shoe will have picked up from the ground and putting it into a context where it might contaminate our food. However, this real tangible danger is lost in the taboo itself, and the taboo has become the major issue rather than the reasoning behind it: If we put brand-new, shop-bought shoes on the dinner table, our reactions tend to be similarly one of disgust. It is the shoe transported out of its proper place and into a food context which breaks the taboo, not the potential germs it introduces into our food.

Risk taboos tend to be associated with the breaking of bodily boundaries, and this is where harm is most naturally introduced to the individual. Bad hygiene practices, rotten food, unsafe sex – these are all inherently dangerous and as such have become the focus of many ritualised dirt taboos. However, excretions that leave the body (menstrual blood, semen, pus, etc.) likewise cross the bodily boundary and are also frequently subject to taboos. This importance of the bodily boundary as the mechanism through which dirt taboos are born also finds its expression universally, and our society is no exception; we have rules surrounding hygiene and food that are ostensibly based on medical knowledge, but we find their expression in ways that go beyond or sometimes even against what we know scientifically to be harmful. Vaccine aversion (Fairhead and Leach, 2012), for example, could be looked at through the fact that injection crosses our bodily boundaries even more directly than oral ingestion. While of course there is much more to be written about vaccine aversion, particularly as part of the coronavirus crisis, a Douglasian reading of vaccines as a contemporary dirt taboo might at least add some interesting insights to our understanding of why there is so much popular resistance to vaccination.

Breaking cultural norms in these areas introduces potential harm to a society. One clear consequence is that that other societies, which may have a different set of cultural norms, are inherently dangerous as a whole. They will have their dirt taboos, which may well be very different in detail than ours, even if the underlying physical dangers like rotten food and disease are similar. Dirt taboos tend to have the universal characteristics as laid out above (breaking boundaries, bodily fluids, introducing disorder, etc.), but how exactly they will manifest, what they mean and what the consequences of breaking them are will be culturally contingent. Therefore, an outsider's actual behaviour will likely break our taboos, and ours theirs. Since dirt taboos tend to be associated with bodily boundaries, other societies' norms and practices in these areas will be most likely to break our taboos. This would be mostly behaviours that concern food intake, sexual practices and norms around cleanliness and hygiene. In this context, it is unsurprising that xenophobic slurs towards other societies often revolve around precisely these traits: The cultural "other" is depicted through the food they eat, their sexual habits, their hygienic practices. As Douglas notes:

> Each culture has its own special risks and problems. To which particular bodily margins its beliefs attribute power depends on what situation the body is mirroring. It seems that our deepest fears and desires take expression with a kind of witty aptness. To understand bodily pollution we should try to argue back from the known dangers of society to the known selection of bodily themes and try to recognise what appositeness is there.
>
> (1966, p. 150)

Douglas theorises that a society can itself become a ritualised metaphor for the body, and in that way breaking cultural boundaries and margins is analogous to breaking the bodily ones. The individual human body is a "conceptual microcosm", in Lupton's words (1999, p. 40), of the body politic. Just as danger is introduced to the individual body through the breaking of bodily boundaries that accompany taboo-breaking behaviour, danger can be introduced to a society as a whole through taboo-breaking behaviour that crosses the political boundaries.

Risk from outsiders is therefore especially salient; illnesses which break through the society's bodily boundary, for example, are due to the norm and taboo-breaking behaviour of other cultures. Taboo-breaking and/or other dangerous behaviour often gets represented through a metaphor that sees society as a body. Thus for example terrorism or drug taking has become a "cancer" within society.

In later work, Douglas tried to more explicitly link her social anthropological work on dirt, purity and risk to contemporary society, showing in particular irritation towards psychological studies on risk perception that seem to work on the assumption of risk perception being based on individual, rational calculations, and that the messy, cultural, political issues surrounding societal and cultural reactions to risk (and who is to blame for things going wrong) are relegated to how people think in scientifically less enlightened times and societies:

> The prevailing idea was that Western advances in knowledge had dissolved a tie that everywhere once used to connect morals and danger: with us morals are soberly enforced by moral persuasion and danger is known by technology; formerly lack of technology allowed the wildest accusations of blame to be hurled right and left and strange spiritual agencies to be invented to cover the cracks in plausibility. Magic and taboo were due to ignorance.
>
> (Douglas, 1992, p. 9)

The book that Douglas wrote with the sociologist Aaron Wildavsky (1983) took her approach to analysing the risk discourses of the environmental movement, where they demonstrate how this discourse positions the polluting outsider (politics, industry, etc.) as the taboo breaker who can then be the object of blame. This identification of the common enemy then provides the environmental community with more social cohesion and purpose that enabled its formation into a stronger social identity. This, however, also had the consequence of deflecting approaches to solving environmental pollution issues through cooperation with industry (and similar others) – it is now a moral issue of *us* versus the morally defiling *them*.

Cultural theory of risk in the time of the pandemic

Infectious disease outbreaks that invade our bodily boundaries, both of the individual and of our society, are therefore probably a particularly suitable risk to be analysed through Douglas' approach. As a generalisation, we can see a lot of international discourse around the coronavirus outbreak but also previous infectious disease outbreaks, centred in some way around dirt and purity taboo–breaking practices of outsiders. The lines of blame are, at the time I am writing this, still shifting and turning, and various discourses scramble to become the mainline interpretation of how we should think about the disease. Who is to blame depends on a number of factors, but the social group or identity we belong to is one very strong variable. The years leading up to 2020 have already seen the emergence of the "culture wars", particularly related to the presidency of Donald Trump in the US (Nagle, 2017) and Brexit in the UK (Koch, 2017), where political identities have become increasingly entrenched and unwilling to see and comprehend each other across their boundaries. Currently, the fault lines of who is to blame (if not for the virus, then for the disproportionally high death rates within these two countries) align fairly clearly along the "culture war" division.

What depressingly seems to have become a wider, global norm, however, seems to be attribution of blame to people of other cultures. Initially at least, in many countries in the global West, the blame (given the origin of the virus) goes to the cultural practices of the Far East, especially China. In those countries, however, a counter-narrative has emerged that attributes Western influence as to blame, either through conspiracy-type constructions that imagine the virus as a laboratory-engineered method of suppressing the Chinese economy, or through the incompetent public health strategies in the US and Europe that allowed these countries to become the global hubs of the infection and thus becoming the main threats to global health. Within the US, a narrative is developing that takes the fact that ethnic minorities have higher death rates and uses this to argue that there must be something in their cultural habits that makes them blameworthy for the spreading of the virus (Beckett, 2020) – again, the racial other is to blame. And so, on it goes – exactly which cultural narratives will eventually emerge from all this is still unclear and will probably be easier to assess in hindsight once the crisis is over, but the rough outlines are already sketched in the sand.

That the racial or national other will end up being at fault for the crisis in some way, however – and that this will of course depend on which nation's narrative we are following – fits a pattern of narratives that has followed infectious disease outbreaks in the past (see e.g. Washer, 2004, on the first SARS outbreak). One obvious clue to this are the very names that we tend to give new infectious diseases: These tend to be named, in popular discourse at least, after their country of origin. The country of origin is often merely assumed, but this does not stop the practice: German measles, Spanish flu, Mexican swine fever, Middle East respiratory syndrome and of course now the popular description of COVID-19 as the Chinese or Wuhan virus. Joffe (1999, p. 25) quotes Sontag in noting that syphilis in the 15th century was known as "the 'French pox' to the English, *morbus Germanicus* to the Parisians, the Naples sickness to the Florentines, the Chinese disease to the Japanese". In this way it's the foreigner's and outsider's unclean habits in breaking dirt taboos that is to blame (and remember, we need someone to blame): For example the (imagined) Chinese cultural trait of eating animals we in our culture would deem unclean to eat, in case of the coronavirus outbreak; or the sexual practices of homosexuals during the HIV epidemic (though they weren't foreigners, they were clearly dirt taboo–breaking outsiders within a heteronormative society, and thus fair game for being blamed); or the dirty unnatural animal husbandry practices of the British for those following the bovine spongiform encephalopathy (BSE) episode from the continent.

Risk and identity

Both the risk society and the cultural theory approaches to risk have identified *blame* as one of the central aspects of how societies respond to risk. For us to understand risk, the potential of future danger, is to try and find meaning in it, and to do that, someone is to blame for this. The role of blame becomes more salient the bigger and more destructive and globalised the risk and when it can fit into culturally contingent narratives that revolve around taboo-breaking.

Noting that academic risk research has tended to concentrate on the two poles of the broadly framed sociological and anthropological theories of risk (as outlined above) and the more narrow, cognitive psychological studies on risk perception, Hélène Joffe (1999) felt that there was a need for work in the "gap" by explaining the "subjective experiences of risk, [and] connecting this experience to broader social factors" (p. 1). She does this by taking social psychological theory, mainly the social representations approach (introduced in Chapter 4) and psycho-dynamic theory. Again, one of the central themes in this elaboration of risk is that of blame, what Joffe calls the "not me – other" response to crisis. As with the running example I used in this chapter, Joffe's elaboration of risk is applied mainly to infectious disease outbreaks, following her wider work on the HIV/AIDS epidemic, but the ambition is broader as a social psychological explanation of social reactions to crisis in general, including wars, earthquakes and so forth.

Rather than directly following Joffe's argument here, I will focus on social identity as a rich explanatory framework for social reactions to risk. First, Joffe introduces studies on identity as a way of understanding the blame response, elaborating this further with a more detailed exposition of social representations theory as a social psychology of how people make sense of new and threatening concepts and situations. Social identity, which I have alluded to in Chapter 2 as a sociological demarcation between religious and scientific communities, has become paradigmatic as a concept in sociological, cultural and social psychological studies (Tajfel, 1981; see Hogg and Abrams, 1988, for a general introduction).

Social identity theory rests first on the insight that groups can form a collective identity very quickly and easily, and these group identifications will have an effect on group members' evaluations and attitudes towards ingroup and outgroup members. Within the so-called minimal group paradigm, research has put research subjects into entirely spurious groups, determined for example by a coin toss (Tajfel, 1982), and then observed group members to engage in behaviour that disadvantages outgroup members and advantages ingroup members.

Second, our evaluation of the differences between groups are accentuated to the extent that clear boundaries are perceived between in- and outgroups when the boundaries are in reality rather blurry (Tajfel and Wilkes, 1963). Hogg and Abrams (1988) use the example of rainbow colours to explain how people pigeonhole themselves and others into discrete groups: While in reality the colours of a rainbow gradually run into each other, we tend to imagine rainbows a set of discrete colours instead, with clear boundaries between, say green and blue. Groups will have particular characteristics, values and norms that define them, some of these by definition (identifying as a father for example will necessitate having a child), while others are more culturally contingent. Identifying with a particular sport, say roller derby, will necessitate participating in that sport but also certain cultural norms that have developed within that community. Identifying with a particular religious community will involve accepting some defining doctrinal tenets of that faith (say, the divinity of Jesus) but also a whole raft of often unspoken, uncodified, cultural norms of behaviour that are consistent with that wider community. Identifying as a scientist involves having the right qualifications but also often a much less clearly defined "scientific frame of mind". And so on.

When we *self-categorise* into a social identity group, any of these ingroup characteristics we have will become accentuated in our self-perception. Once we identify as, say a scientist, our "scientific frame of mind", even if never clearly defined, becomes part of who we think we are, and any of our actions, thoughts or behaviours get interpreted as exemplifying this. Other members of the ingroup get a similar treatment: Behaviour that we might otherwise classify as unscientific or at best unrelated to the science identity get understood as according to it.

The outgroup member by contrast will not get this treatment, on the contrary, and of their behaviour that could be construed as according to ingroup norms will be downplayed and reinterpreted as not really ingroup behaviour. Our ingroup characteristics, norms and values will tend to be positive ones, or at least categorised within the ingroup value system as positive, while outgroup characteristics will tend to be negative. Self-categorising ourselves within our identity ingroup and sharply delineating ourselves from the outgroup(s) will enhance our self-esteem.

Social identity theory maps on to the discriminatory and blaming behaviour displayed during infectious disease pandemics. The blame, as shown, lies with the other. As Douglas has theorised, this will be due to taboo, boundary-breaking behaviour of outsiders; reconceived through social identity theory, we can see that it is the ingroup's own characteristics that are being evaluated as positive and the outgroup behaviour as negative. Douglas' theorisation, however, adds the extra explanation of what kind of ingroup and outgroup characteristics will lead to the discriminatory, boundary policing we find in situations of crises like a pandemic.

On the wider sociological perspective of the risk society, as introduced earlier in the chapter, we can also understand the types of blame behaviour that Beck described with respect to the large globalised, late modern risks like climate change. The blame response attaches not just to people of other nationalities or cultures but whole groups within our society which are delineated from us, the ingroup, by much less than the tightly knit cultural binds suggested by a Douglasian reading: Social identifications – as well as social categorisation of outgroups – can occur with minimal differentiation. The blame for the pandemic then attaches to the traditional outgroups of foreigners or clear subcultures within our society, but they also attach to the scientists, governments and elites of our society. If we are part of the marginalised culture ourselves, the blame can go to the elite quite easily, and even the perception of being in the marginalised culture is enough for discriminatory behaviour to appear, as Joffe has described for example in the various evaluations of where the HIV virus came from.

Here is maybe another link to conspiracy theories, which tend to be part of marginalised culture. This is often very much self-consciously, so as Robertson's (2016) study showed, conspiracy theorists described themselves as "conspiracy nutters", demonstrating that theirs is a stigmatised identity that needs managing (as in Goffman's influential 1963 argument), though it bears noting that Harambam and Aupers (2017) did not find this discourse. Generally, though, the targets of conspiracy theories appear to be the elite, even if only often the imagined elite, who are to blame for social crisis.

Risk and apocalypse

The three broadly sociological theoretical strands I have introduced in this chapter are in my view complementary rather than really different ways of explaining the social meaning of risk. The concluding part of this chapter will attempt to bring these strands together and relate them beyond my running example of the coronavirus pandemic and relate them to the wider apocalyptic movements and imaginaries I outlined in previous chapters. I will then attempt to construct a tentative answer to the observation with which I began the chapter, that is the use of "risk" terminology as a curiously contemporary approach to the apocalyptic.

The way that we understand the catastrophic future is intimately bound up, within contemporary discourse, with risk: The future is knowable, to an extent at least, and though we frame the future in terms of probabilities, knowing these probabilities lets us control the future. In this sense, risk has replaced fate as our imaginary of the future. But while risk is often conceived within the technical and psychological literature as a calculation and therefore a way for us to control the future (or to attribute biases when the calculations are wrong), as we have seen on a broader sociological look, the issue of risk has become less about probabilities and control and instead one of how societies react to danger. When risk is no longer about whether we can control it, the pertinent issue becomes one of who is to blame. It is unsurprising that all three of the sociological approaches I have presented here focus on blame, either as a constituent part or even as the main explanatory motivation for the theory.

In any case, risk is not a clearly defined concept – it may be within any of the disciplines that work with it, but as a wider, contemporary concept in everyday language it has absorbed all these different aspects, of being deterministic in some senses but avoidable in others, of being calculable in one sense and uncontrollable in another. Risk is a social representation, a boundary object that eases communication between social groups (in this case, including different academic tribes) but leaves each group's general understanding of the concept intact even when interacting with a group that has a different idea of what risk actually is.

However, risk, particularly when we talk about the "existential", globalised risks of late modern society, appears to me particularly comparable with the "apocalyptic", which as we have seen is similarly polysemic and interpretatively flexible. The apocalyptic can mean the certain end, the probable end, a possible but avoidable end; it can mean the end in a nihilistic sense or the end in the sense of it being the start of a new era. While I would hesitate before naively equating the apocalyptic with the existential risks of today, my argument would be that they perform at least comparable work within our narratives and imaginaries of the future; and, as with religion and science, any boundaries between them are not clearly defined or immediately obvious. If this is the case, then we can also try to use some of the sociological theory that has been developed about risk to analyse the apocalyptic in a wider sense.

Any apocalyptic movement is characterised by a strong identity. As research in social identity theory and the minimal group paradigm has shown, the formation of group identities strong enough for in- and outgroup projections and stereotyping to occur can happen very easily and on minimal actual differences between the groups, and therefore any religious group that has physical meetings, common beliefs and friendship networks will form a social identity. With that identity then come positive ingroup evaluation, an increased (self-categorised) conformity within the community of ingroup norms and values, and disparaging evaluation of the outgroup(s) (Burris and Jackson, 2000; Bloom, Arikan and Courtemanche, 2015). In this respect religious communities and apocalyptic movements will be no different than, say allegiance to a particular football club or Oxford college.

Prophetic pronouncements act as the binding belief characterising ingroup norms and values. Some of these, maybe particularly so the apocalyptic prophecies, can become group defining, in the same way that being a hockey player is indispensable for the hockey player social identity, while some other prophecies or behavioural norms can be subject to change, such as the behavioural norms or values expected within the hockey community. In this sense, there is a link to some of the social psychological work on social identity and the hardcore versus protective belt of belief statements introduced in the previous chapter. Social identity theorists talk about "prototypicality" (Waldzus et al., 2004), which then may map onto Lakatos' indispensable group characteristics.

Once the social identity is defined, with apocalyptic beliefs that are defining or prototypical to the group, and a protective belt of social norms and values that relate to some of the dirt taboos as analysed by Douglas, then we can arrive at a more detailed understanding of how apocalyptic movements react to outsiders, and how outsiders are seen as introducing risks to the group by breaking of dirt norms and taboos. The apocalyptic, the "existential" risk, is something that outgroup members impose onto the world and thus also onto the ingroup, through their dirty and disgusting ingroup taboo-breaking habits.

There are then connections here between the different ways in which the apocalyptic has manifested itself: The traditionally religious movements that see the apocalypse as being caused in some way through outsiders, even if this may introduce a slight contradiction within the belief as the ultimate conduit of the apocalypse is God or fate (She destroys the world, after all). Revelation (or the various denominational interpretations of Revelation) is full of enemies, outsiders who cause the events to unfold, and even though they are fulfilling God's plan for the world, they are still morally responsible for the evil they cause. Some of these enemies are supernatural, but mostly they are the "traditional" outsiders of foreigners (the hordes from the East, or the armies of Gog and Magog, or the unbelievers and doubters, the outgroups from within our society who distinguish themselves by not taking the righteous path. In these and similar prophesised apocalypses, it is the outsiders who cause the end, through their amoral, disgusting and taboo-breaking behaviour.

Similarly we will find discourses of outsiders and taboo breakers in other religious and non-religious apocalyptic belief systems: The Xhosa cattle massacre had the colonial British as outsiders; Festinger's UFO cult had the non-believers as the outsiders (who would cause the great flood through their amoral behaviour); the great humanist revolutions had their outsiders who precipitated the events that led to the cataclysm of the revolution (remember that capitalism is the seed of its own destruction, that is the fault for the revolution does not really lie with the revolutionary, but the class they overthrow), and it is the same outsiders who are to blame when the revolution has not been able to (yet) deliver its utopia.

And it is apparent with the "existential" risks of modern society, the moral fault lies with those who break the taboo and purity rules that are cornerstones of environmental identities, such as the dirty and polluting behaviour of morally deficient chemical and agricultural industries (Douglas and Wildavsky, 1983). It is the breaking of environmental taboos by outsiders that endangers the whole community and introduces the apocalyptic risks. Just to reiterate that this is not to say that un-environmental behaviour doesn't cause risk to the now-global community, the science is clear, but the risk narratives that surround these taboo-breaking behaviours are comparable in their social function to the wider apocalyptic narratives and can thus be analysed using the same social theoretical tools. It does, however, show how environmentalism can become an identity, or rather one of several identities, as the environmental movement splinters denominationally into several groups with sometimes conflicting ideas of what the root cause of the problem is, and what exactly the taboos are whose breaking causes the most risk to society.

Chapter 7

Utopias

Millennium, utopia and revolution

The millennial kingdom of God that will appear after the apocalypse can be seen as an instance, or a precursor, of the concept of utopia (Kumar, 1991). Commonly understood as an imaginary perfect society named after the influential genre-setting work of Thomas More, it is part of a long tradition of millennial dreams about what the world could or should be. As one of the various Blumenbergian mythic traditions that appear to have survived the switch from religious to secular society (demonstrating that this shift has never been as complete anyway), utopia and the millennium are linked through being fantasies about what our society could potentially be, usually in the future, but for various reasons is not yet realised. These imaginaries of a better life can be divided into various strands representing different traditions; some authors such as Davis (1993) divide utopias into fantasies such as the land of Cockaigne (the German *Schlaraffenland*), where every desire is fulfilled; Arkadias, where our desires are more moderate; and "moral common-wealths" where collective problems are solved "not by increasing the range or quantity of satisfaction available, but by a personal limitation of appetite" (p. 31).

Kashima and Fernando (2020) argue that the reason why millennias, utopias, Arkadias and other imaginaries of the perfect society so persistently show up throughout history in most cultures indicates that they in some way perform important social functions. Citing Levitas (1990), they list these as criticism, change and compensation – a utopia (or millennium) acts as a mirror to criti-cise current society, it directs action towards a more or less clearly formulated goal and it "compensates for the troubles and mayhems of worldly affairs in an escapist dream" (p. 102). Thus, they argue, utopia acts as a mode of "collective self-regulation".

In any case, millennias and utopias (as with most concepts these two are not entirely clearly defined, and thus the boundaries are blurry and fixed by conven-tion rather than content) are an integral part of imaginaries about the end of the world – as with our own personal deaths, thoughts about the death of society often focus about what comes after. We can go to heaven, hell or nowhere at all; simi-larly, we can envisage utopias, dystopias or nothing at all after the apocalypse.

However, that the end of the world is often not really the end but rather the transformation of our current imperfect world into a new, perfect one is one of the defining features of the apocalyptic imaginary. The apocalypse as it unfolds in Revelation recounts a 1,000-year period of peace after the fall of Babylon, during which Satan is confined to the abyss. This millennium is not the final end state, since at the end of 1,000 years Satan will be released again and defeated in a final battle, after which judgement day will divide all souls into an eternal lake of fire or to live eternally in the city of heaven, where there will be no pain, tears or death. The chronology of these events is very much disputed (see Leithart, 2018), as is the extent to which St John's words are to be taken literally, figuratively, allegorically or metaphorically. In some interpretations the millennium is yet to come, whereas in others it has already started and (with the establishment of the Catholic Church) we are living through it right now.

For the Christian traditions that have collectively formed our (Western) cultural background, whichever precise interpretation of Revelation we take, it is clear that, in Leithart's words, "Revelation *is* about an end. It is not about *the* end of the world, it does, however, predict the end of *a* world" (Leithart, 2018, p. 24). If it is the end a world, that also means that there is a world after that, and while Revelation is notoriously vague and complicated, this aspect appears clear at least. Christopher Rowland (cited in Leithart, 2018, p. 57) discerns an "apocalyptic worldview" that characterises Christian eschatology in Revelation and the other, smaller biblical apocalyptic texts as a dualistic cosmology. Leithart notes that this worldview includes a dualistic cosmology that divides things into clear good and evil – people, places and temporal worlds, where the wicked earthly world is contrasted with the heavenly world to come. It also includes "dissatisfaction with the world as it is", meaning that apocalypticism is "ultimately hopeful because God is both sovereign and interested. He will rescue the righteous and punish the wicked" (Leithart, 2018, p. 57).

Though they are terms that are derived from the 1,000-year period of peace in Revelation, "millennium" and "millennialism" have in academic literature widened to refer to the expectation of a heavenly time of peace and justice after the end of the world, whichever way the rather complex structure of Revelation is interpreted, or whether we are believers at all (see Wessinger, 2011b).

Millennialism in the form of the imagination of a better or perfect society found its way through the "millennial communism" that Cohn described in the medieval apocalyptic movements he described, and which can be found in more contemporary examples as well. Either as a way of realising the real "city of God" or the "new Jerusalem" in this world, as the Münster Anabaptist thought, for example, or as a precursor society that makes its members lead a virtuous life in preparation for the day, millennial visions for the perfect society often link to communitarian visions of equality (within group members at least) and otherwise adherence to a virtuous life.

As a tradition possibly better viewed as parallel to millennialism, the imagination of a better or even perfectly structured society has found its expression in more worldly orientated tracts on political philosophy, with of course Plato's *Republic* being an early and the most influential example. This way of imagining a perfect world was clearly practically orientated, with a view towards setting examples and thought experiments on how current polities could and should be run.

The word "Utopia" itself was coined by Thomas More in his 16th-century work of that title (1989 [1516]); the name "utopia" is a play on words and could be interpreted as either "nowhere" (outopia) or "the good place" (eutopia; Levitas, 1990, p. 2). More's book set a standard of a new genre of utopias or imagined perfect or near-perfect societies that has been imitated heavily ever since and especially within early modern literature. Though More cites and draws heavily from Plato's political philosophy, Kumar (1991) argues that More's writing was innovative enough and influential enough in its own right to be regarded as the real origin of the utopian imaginary, rather than Plato. As distinct from Plato, More imagines his perfect society as a fictional place, with the book presented as a travel narrative. It is this fictional element that presents the perfect society as a dissociated social critique, where criticisms of the current society could be projected as innocent fantasy.

This pattern was followed by More's many imitators and followers, the utopian society was physically usually just out of reach of our current knowledge of geography, moving further away as exploration mapped more of the world. More's utopia, being written at the time of the exploration of the Americas, could still plausibly be situated within the Atlantic; by the time Bacon wrote his *New Atlantis* (Bruce, 1999), the Americas had been mostly mapped and European explorers were starting to chart the Pacific, which is where Bacon placed his utopian island. Thereafter, utopias could be set, for example, on alien planets (Kumar, 1991, p. 58) – always just enough out of reach to remain plausible. With the utopia as a plausible but at the same time clearly fictional narrative, it could function as the easily deniable social critique. A similar function is also performed by utopias (and dystopias) in science fiction, where imaginaries of future societies could become a social critique without maybe always being spelled out explicitly as such (Fitting, 2010).

Davis (1993) distinguishes between the informal millennium and the formality of (early modern British) utopia, with utopias being formal and law based (Avilés, 2000) in line with the general social values of the time, whereas the millennial informality "meant embracing the dissolution of all human social contrivances, the destruction of all formalities" (Davis, 1993, p. 21). There are differences between these utopian imaginaries and the millennial ones described earlier in that for the millennium, the specifics of the perfect millennial post-apocalyptic societies are rarely described in detail. It's enough to say that God says everyone will be happy, that we will eventually find peace and prosperity, and how this exactly is going to happen is not worried about in scripture, and it won't be our responsibility to enact through codes of law or behaviour. By contrast, utopias, certainly those of the Plato-More tradition, provide clear blueprints for how they will work. However, this is not a clear delineation. Millennial movements that try to institute the city of God here on earth always must end up having to face the practical details on how these societies should be organised. Even if they were not written down as aspiration at the start, they end up as societies governed by strict codified rules or behavioural norms.

These utopian narratives also differ from the millennial hopeful stories that motivated apocalyptic and millennial movements by being clearly marked as fictional and thus never to be destined as such, unless of course clear political action follows to make this happen. But as outlined above, inspiring clear political action was not really always the point of these utopias; instead they were mirrors on contemporary society pointing out areas where it is wrong or could do better. In any case, the agent of change in ushering in a better society would need to be us as political players and societies rather than the preordained future envisaged for us in scripture. This could be seen as part of the general shift that accompanied the change in political, scientific and religious imagination with early modernity and then continued through the enlightenment: The scientific and social revolution of these times ushered in an increased secularisation that, while not abandoning God or scripture, increasingly bracketed it out from our lives and placed the responsibility for our world onto ourselves and away from God. However, that this might be a slightly naïve interpretation on what happened, and the move to secularism, was a more complicated process (Taylor, 2009, ch. 2). It appears, then, that on the topic of millennialism/utopianism we have also not as clearly embraced a modern secularism. While More-type utopias were in a sense a newly invented genre, as noted, imaginaries of perfect societies have been part of European culture since Plato. Conversely, millennial imaginaries on future perfect societies – ordained and predestined by God and scripture – have also happily continued, and even thrived, throughout early modernity and after.

The enlightenment thinkers that influenced the new humanist thinking which eventually provided part of the intellectual background to the American and particularly the French Revolutions, were also often engaged in utopian writing. For example Kumar (1991, p. 57) discusses Diderot's "Supplement to Bougainville's Voyage" and de Sade's utopias based on sexual liberation. While Kumar notes that these utopian writings "followed rather than led eighteenth-century thought" (p. 58), the utopian and millennial connection between the hope for an imminent better society to be achieved after a period of struggle and tribulation, and the revolutionary air of the times seems clear. Armed with enlightenment visions of how the good and egalitarian society can be plausibly achieved, as well as with the maybe unconsciously transmitted millennial hope of a society influenced by centuries of Christian tradition, the revolutionaries could find philosophical justifications underwritten by the will of God for the overthrow of their contemporary, clearly non-perfect societies. The link between the utopian ideal and revolutions (that after all, aim to bring it about), means that there is only a short conceptual leap into identifying utopia as a defining characteristic of the totalitarianism that often follows millennial revolutions (Gray, 2007).

The link between the humanist revolutions and millennialism have been noted by apocalyptic scholars as well (see particularly Landes, 2011). Though they are not formal utopias, Barkun (1974) for example sees the doctrines of "inevitable progress and the total redemption of a corrupt society" (p. 28) of writers such as Rousseau and Condorcet as being part of the millennial tradition. Part of this revolutionary philosophy was a "utopianism which sought the final and ultimate attainment, by force if necessary, of totally harmonious human society" (p. 29).

The humanist millennium/utopia carried on with the development of social- ist ideas in the 19th century, most influentially through the writings of Marx and Engels. Though Marx was apparently dismissive of "cookshop" (quoted in Kumar, 1991, p. 61) utopianism, his post-capitalist society had all the hallmarks of the utopian, even if it lacked the detailed prescriptiveness of More and his imitators. In any case, the lack of detailedness in Marx's thought did not stop the communist revolutions trying to enact the perfect communist society, just as the events of the French Revolution focussed more on the overthrow of the corrupt society than thinking through in clear detail what its replacement should look like, since all the carefully drafted enlightenment utopias foundered at the encounter with the cold, difficult and messy reality of having to actually implement a perfect society.

The tradition of humanist revolution and transition from the corrupt, imperfect society to the utopian imagined one, whether only foreseen as a vague template or whether described in detail through formal utopian texts, was often however also accompanied with religious or quasi-religious narratives that borrowed heavily from the apocalyptic tradition, with saviour and messianic figures and secularised proph- ets, for example. These can be seen in the personality cults of the strongmen who will force society into a new path of righteousness. But there is also a recasting of the revolutionary struggle not just as the hope for a world that is better than what we had before but as a clear struggle between an almost biblical good and an unredeemable evil. For the French revolutionary, this was the aristocrat who became synonymous with the evil of the ancien régime, for the socialist and communist revolutionary the capitalist who suppressed the workers and who morphed into the caricatured evil capitalist Jew and then just the evil Jew of Germany's national socialists.

Theorists and critics of utopia

With the aftermath of the Second World War, with the atrocities committed through Nazi millennialism and with the cold war raging against the communist millennial regimes in China, the Soviet Union and elsewhere, an intellectual current devel- oped within Western Europe and North America that not just actively questioned the utility of utopian world building as a political project but also started to see it as part of an ideological danger that needs active combating. That utopian and millennial imaginaries should become to be seen as dangers to be overcome is clearly understandable in the light of the atrocities the war and the ideological rivalry with communism.

Mannheim

Karl Mannheim wrote his most famous book, *Ideology and Utopia* (2015 [1936]) at a time when the late 1920s German society was starting to fall apart partly as the consequence of the country's defeat in the First World War and the global financial crash that Germany had felt particularly harshly. The Weimar Republic was struggling to retain authority and a warring communist and an emerging fascist ideology battling against each other, with the fascist eventually winning. This period marked the end of the relatively brief and only period of peace that Mannheim lived through (Shils, 1974). This milieu, as Shils suspects, coloured Mannheim's outlook and sociological thinking about utopia and how it relates to ideology (see also Wirth's 2015 [1936] preface to the English edition).

One of Mannheim's insights was that Utopia as the imagination of a better place is not readily distinguishable from other political projects put forward by opposition parties as a suggestion of how things could be done better (Geoghegan, 2004). This view can be challenged as widening the definition of utopia so much as to be unproductive, and for this reason Mannheim's work does not feature in detail in Kumar's introductory text on utopias other than to say that his conception, and the dominance it has achieved within sociology, is not particularly helpful – they are, in Kumar's words, "properly to be regarded as expressions of those systems of thought which merely anticipate reality, rather than deny it" (p. 92).

As such, utopia can be linked to ideology as they are both exercises in "reality transcendence" (Geoghegan, 2004), with the main difference being that utopias are oriented towards the future whereas ideologies are oriented towards the past, and utopia can morph into ideology when normative political visions are actualised and then become part of past political life, in a process that Geoghegan identified as partly the influence of Hegel and Marx's dialectical view of history on Mannheim. Both utopia and ideology are distortions of reality, they are imaginaries of what could or should be rather than what is, and "what in a given case appears as utopian, and what is ideological, is dependent, essentially, on the stage and degree of reality to which one applies this standard" (Mannheim, 2015 [1936], p. 176). The term "utopia" can then be used pejoratively as anything that from the point of view of the existing social order can never be realised, ideology can be used pejoratively as a defence by the utopians who see utopia instead as a "socially grounded insight into the defensive ideological illusions of the dominant strata" (Geoghegan, 2004, p. 125).

Mannheim's conceptions of utopia and ideology as transcendent realities links to his general quest to develop a general sociology of knowledge. But his treatment of ideologies and utopias as transcending reality raises the issue of what really is real (Levitas, 1990, p. 89, recounting Ricoeur's criticism of Mannheim): "Mannheim does not perceive that social reality is always and essentially socially mediated, rendering the opposition between ideas and reality . . . untenable" (p. 89). With *Ideology and Utopia* as part of his developing sociology of knowledge, Mannheim's argument can be seen as an exploration of "the implications of the fact that all knowledge, and particularly all knowledge of the social world, is partial, selective, and dependent upon the social location of the observer" (p. 93). This, as Levitas argues, links Mannheim as an early formulation of the "problem of relativism", the sociological version of the fundamental epistemological questions about how we can evaluate truth claims – if all knowledge is arrived at from the position of the observer, "how can we evaluate competing versions of the state of the world?" (p. 94).

His link of utopian thought with ideology brought his analysis in line with the wider, often Marxist contemporary treatments and critiques of ideology, but with the addition of adding similar critiques to Marxist normative political thought. For this reason, Mannheim can be considered as influenced and sympathetic to many aspects of mid-20th-century Marxist philosophies but also explains some of the negative reviews he received from members of the Frankfurt school (Kettler, Meja and Stehr, 1990).

Dystopias

The idea that utopian thought of the classical "blueprint" tradition is too rigid and inflexible for them to function is one of the objections that can be made to utopia (Moore, 1966). If all social relations and problems already have worked out solutions formulated in a clear blueprint, then that society will be ill-equipped to react to new situations (or to iron out unforeseen problems inherent in the blueprint). As real-life utopian societies have found out regularly, a rigid adherence to the founder's ideals is often a reason for failure, and the rare utopian societies that have survived in the longer term are those that managed to adapt and become flexible in their interpretation of the rules (Kumar, 1991).

These societies can then turn into a dystopia. Dystopias are often seen as merely the reverse or opposite of utopia, the imaginary of the imperfect society, but in some ways a dystopia is the other side of the utopian coin, or a consequence of a dysfunctional, inflexible utopia or one that was not properly thought through. Unlike utopia, which is not just a literary genre but can, as per Mannheim, also be seen as a normative political project on how society can be better run, dystopia tends to be a term applied more purely to literary works, since of course a political project would not intentionally project a worse world as an ideal to strive for. The literary dystopia often results from attempts at building a perfect society that has turned sour, and in this it is somewhat in line with the failed utopian societies of real life.

Some of the famous dystopian novels from the mid-20th century apply projections into the future of utopian ideals gone wrong. For example George Orwell's novels *Animal Farm* and *Nineteen Eighty-Four* (Orwell, 1946, 1949) famously satirised the communist and fascist societal "ideal" societies and how they can easily morph into nightmares. Imagining how fantastic scientific advances can be used in future utopian societies to make life easier is an established trope of science fiction. Some of these imaginaries remain proper utopias, but more often this genre of science fiction also imagines how things can turn sour, at least depending on the reader's viewpoint, such as Aldous Huxley's *Brave New World* (1932). Thus, as Gordin, Tilley and Prakash (2010) argue, dystopia is not so much the opposite of utopia but rather a utopia gone wrong (or a utopia for only one section of society).

Anti-utopia as a political project

By the 1970s, utopia, ideology and Mannheim himself have all fallen out of fashion in sociology (Kumar, 2006). The second half of the 20th century saw several influential philosophers such as Karl Popper and Friedrich Hayek formulating explicit arguments against utopia, again like Mannheim or Orwell most likely as a direct reaction against the utopian projects that plunged the world into war and genocide (see also Gray's 2007 argument on utopia and totalitarianism). Popper argued that a Hegelian-Marxist style view of history fundamentally misconstrues how history should be understood. There is no clear direction that history is fated to take, and therefore Marxist-style projections of us eventually arriving through the downfall of capitalist society at a perfect, final society is misguided and unrealistic (Popper, 2011 [1945], 1963, ch. 18).

Leaning on his own philosophy of science, Popper instead proposes a different view on how political projects should learn from history and aim to run society. Science, as per Popper, learns and advances through a continual process of proposing hypotheses, testing them to destruction, and when they fail, amending them, continuously with something better. Like this, science makes continuous, piecemeal advances: It never finds out final truths; the only finality it finds is about what is definitely not true. Popper's political philosophy then likewise envisages not an unchanging blueprint, which will sooner or later run into trouble with reality, but instead a project that continuously makes small changes and adjustments that are able to react and deal with unforeseen events or problems. There will never be a perfect society, just as we will never alight on the scientific truth, however, just like science will slowly approach an approximation of the truth through a process of conjectures and refutations, so a well-run society will continuously approach the never-realised perfect society. Being non-ideological (and thus suspicious about utopian) as a political philosophy has also become more generally a feature commonly associated with conservatism (Hamilton, 2020).

With the end of the cold war and the collapse of communist regimes in Eastern Europe, the victory has been declared as meaning the end of ideology, with Francis Fukuyama's influential (1992) "end of history" thesis proposing to rehabilitate the Hegelian historicism that Popper criticised and argue for a progression of history with a more or less clear endpoint of liberal democracy as the conclusion towards which humankind will develop. This is not an end to ideology though, and Fukuyama (1995) also makes clear that it has often been misunderstood by taking his normative argument as an empirical one.

And of course, history continued being made, and ideology is still very much with us. Conservative parties that had previously been perceived, and often perceived themselves as non-ideological, both in the US or the UK appear to have in recent years ideological and even utopian projects. In the UK this centred around Brexit and the future millennial imaginaries of a new and better society it inspired in its proponents. In the US, the tea party and later the Trump movements also had clear indications of ideological and even utopian strands (with the utopian partly being supplied by the millennial evangelical forces that gradually took over conservative US politics after Reagan). And both have been extensively analysed as showing the totalitarian leanings that authors such as Gray (2007) associate with utopian worldviews.

This conservative embrace of ideology and utopia should not have come as a complete surprise, however. The disavowal of utopia/ideology as a political process that animated Popper and other conservative thinkers may have had its methodological merits but was itself not divorceable from ideology: The whole background of non-ideology must be an ideology itself; any interpretation of the world and how it works or should work needs to be made from within a particular set of background views and assumptions that in the end make up an ideology for all intents and purposes. In Popper's own philosophy of science terms, all observations are themselves theory-laden; in a similar vein, I would argue that all political observations and the interventions that follow are ideology laden. This is clearly a very broad sense of ideology in that it is an inescapable part of how we understand the world, that is we must have a background set of views and assumptions in order to interpret any new piece of information. But this links back to the Mannheimian view that all knowledge is dependent on the observer.

In a narrower sense, however, ideology informs the utopian visions that we have of the future, even if these are not acknowledged as utopian (or ideological). As with Mannheim's insight that any imagined future way of doing politics is in some way utopian (or ideological), so do Popper and Fukuyama (and any other conservative political thought) imagine a utopian situation. Fukuyama was quite explicit about this in that his view of liberal democracy can be seen as the end-point for human social development.

Apocalypse, techno-utopia and (science) fiction

Bruce (1999) leans on McKeon (1987) to argue that the early modern utopia can be seen as a "precursor" to the novel as a literary form. The utopia is "a text which engages very directly with some of the epistemological and social questions which the novel has recently been understood to confront and formulate" (p. xvii). In particular, Defoe's *Robinson Crusoe* has been labelled as one of the earliest modern novels and as coming out of the utopian tradition that uses a traveller narrative as a means of imagining the building of a new society (James, 1993). More broadly, the link between utopia/dystopia as an imaginary of the ideal or gone-awry future and fiction is probably easy to see. In particular, the speculative fiction genres such as the Robinson-style utopian travel narrative, but also the fantasy and science fiction genres that have developed in some sense out of the development of the modern novel, invoke imaginary societies and imaginary ways of living.

Next to the imagined future of utopia and/or dystopia, modern fiction also evokes the end, or the apocalyptic as the transition between the current and the imagined society. The apocalyptic as a central part of modern literature has been explored famously by Frank Kermode (1967). Kermode argues that, in Wagar's (1982) words, "modern fiction and drama is nothing less than a secularized and immanentized version of ancient apocalyptic prophecy" (p. 10).

While Kermode has written mostly with mainstream literature in mind, arguing that "the end" is a permanently immanent feature of the personal narrative within all modern fiction, the apocalypse is often a more direct and literal feature within literature. The utopian millennium and the apocalypse that brings it about come into play particularly often in science fiction, as imaginaries of the future.

The science fictional imaginary can often be fairly straightforwardly used as a utopian exploration of a better society in a fairly direct lineage to the early modern utopia genre initiated by More, with for example the fictional *Star Trek* universe portraying a humanity that has transcended internal conflicts, this society having been achieved only after the apocalypse of a worldwide nuclear war (Gonzalez, 2015). Like the criticism function of utopia in general, as noted by Kashima and Fernando (2020, see above), science fiction utopias often reflect the social issues of the authors' times and the inclusion of historical reflections on how the utopia was achieved can aid the reader in imagining what changes can be made to change their own society (Williams, 1978). Since utopia and dystopia are not really opposites (Gordin, Tilley and Prakash, 2010), dystopian fiction can in this respect be argued to fulfil a similar role as utopia. The imagination of a considerably worse society can be taken to be just as much an ideological exploration of what is wrong with our current lives; the dystopia is often an extrapolation of one or more undesirable aspects of our current society taken to the literary extreme.

But fiction can also invoke the apocalypse directly, through the use of the radical break that the apocalypse represents as a device that can bring about the transformation to the utopia or the dystopia that it wants to explore. What exactly the apocalypse is and what it consists of varies, but the story is rarely about the apocalypse itself, but the society that follows after.

In line with Wagar (who concentrates on literary science fiction), Shapiro (2002, who concentrates on cinematic nuclear bombs) and others, the analysis of apocalyptic fiction shows that the post-apocalyptic dystopia is a clear narrative trend. These books describe many notable examples, many such as Walter Miller's (1993 [1959]) classic post-apocalyptic novel *A Canticle for Leibowitz*, which imagines the world after nuclear apocalypse and its slow rebuilding anew, do not describe a utopia by any means but still imagine the apocalyptic cataclysm as a new beginning rather than an end. Other post-nuclear fictional dystopian landscapes, like the brutal wastelands of the *Mad Max* movies follow a similar pattern, the dystopian future precipitated by the apocalypse portraying a "genuinely nasty worldview, and a politics of absolute nihilism" (Hassler-Forest, 2017,

p. 301) but also invite more hopeful visions of redemption and the rebuilding of communities.

The apocalyptic narrative tropes that I have been arguing appear in our everyday interpretation of (secularised) apocalyptic events are very much in evidence in fictional treatments of apocalypse as well. The post-apocalyptic millennium, both in its positive utopian sense, and also in its more satirical dystopian one, is clearly discernible and has been the treatment of a variety of academic work (e.g. Jameson, 2005; Horn, 2014). To this, however, we can add apocalyptic narratives tropes more widely – the cosmic battle between good and evil that is part of the apocalyptic theodicy as explored by O'Leary (1994) is clearly present in fictional treatments of apocalypses and post-apocalypses.

While science fiction and its connection with the millennial aspect of the apocalyptic has been looked at extensively as outlined above, the cosmic battle element of the apocalypse can be seen maybe more widely within other genres of speculative fiction, particularly the fantasy genre. The world of Tolkien's *Lord of the Rings*, one of the foundational texts of this genre, takes this battle into the foreground. The apocalyptic in the *Lord of the Rings* is not so much evident in any particularly cataclysmic or catastrophic event that brings about the end of times and a post-apocalyptic new beginning. However, the post-apocalyptic, hopeful look into a better future is there: After the final vanquishing of the "beast", a new age of peace and prosperity begins. Thus for example the literary world constructed by Tolkien can be interpreted as apocalyptic and millennial as well (Treloar, 1988; Fontenot, 2019).

While the previous section looked at (mostly science) fiction as a way to construct an imaginary society of the future, and the sections above that looked at political and philosophical thought about how societies in the future could look, one maybe slightly overlooked aspect of utopian is the scientific and technological utopia but not imagined in fiction so much as in political thought. With this in mind, Kashima and Fernando (2020) added *science* to Davis' (1983) utopian prototypes. This techno-utopianism manifests itself in the often caricatured and possibly stereotyped thought that there exist, or that there will exist, technological solutions to our problems, including social problems.

The optimistic scenario envisaged by Asimov (1979), discussed in Chapter 4, which outlined a succession of (mostly but not exclusively natural) civilisation-ending catastrophes is one such example, and it is possibly not an accident that this vision was espoused by a popular science writer who had a much more successful career as a science fiction writer. A catastrophe of our sun reaching the end of its lifespan, for example, was for Asimov something that should not worry us particularly because by the time it happens, we will have found a solution to it.

Asimov was writing within a particularly technologically optimistic time, the rapid growth of scientific and technological knowledge during the military-industrial-scientific complex emerging in early cold war years; the "atomic age" promised technological solutions to the social problems that have been dogging the earlier half of that century, the fears over nuclear war, which were not shared by everyone at least at first, notwithstanding. This technological utopia could, like other utopias as discussed above, easily turn into dystopia in literary imagination; it was partly also this technological optimism which was parodied by Huxley's (1932) *Brave New World*.

However, even outside of the context of the atomic age or any other explicit affirmations of how technology will feature in the coming social utopias, or maybe dystopias, there is an inherent utopian project hidden within the very idea of technological progress that doesn't need to be explicitly formulated at all: If it were not a means for making the world better in some ways, then the notion of technological progress would make no sense. And this would even apply to dystopian technological progress such as nuclear, chemical and biological weapons research, as this is rightly or wrongly justified by its proponents as furthering a good cause in some way, for example as a deterrent (e.g. Balmer, 2002). In this, the idea of inherent utopia in technological progress is similar to the implicit utopias hidden in most political thought, as realised by Mannheim, because any normative political philosophy necessarily entails envisioning a better way of running society. As Paskins (2020, p. 84) recently put it, "Scientific practices may be considered utopian in a number of senses: they can embody hope, or values and ways of relation which challenge habitual ways of acting".

Partly, as Paskins observes, the scientific/technological utopia can be argued as arising out of the extrapolation of scientific ideals to political life – the Popperian version of rational scientific tinkering through conjectures and refutations mimics how society should function at large; despite Popper's anti-blueprint utopia, as I noted above, his version of the ideal future society is utopian itself – and it looks a lot like science. Similarly, the rational reconstructions of how science as a social endeavour reaches progress as described by other philosophers and sociologists of science can be seen as examples for society at large to follow; as scientists are evidently successful, so society will be successful if it emulates what scientists are doing. Paskins discusses the work of Polanyi in this context, who imagines the scientific community as a "republic of science" (Polanyi, 1962).

Another area where science, science fiction and utopia meet is the environmental utopia (Garforth, 2018). Amid the gloomy predictions of coming environmental disaster, hopes and aspirations have been attached to how society can rebuild a more environmentally sound way, looking back to pre-industrial, Edenic pasts as a model for a better societal relationship with nature and forward into how the coming environmental apocalypse will sweep away the destructive capitalist ideologies responsible for it. These utopian ideals are expressed in radical green movements attached for example to Arne Naess' "deep ecology" philosophy (Drengson, 1995), as well as in some fiction and science fiction narratives. Some of these themselves influenced the green movement, such as Callenbach's (1975) *Ecotopia*, which in analogy to the More utopia genre was also written as the travel experience of an outsider to the imagined utopian state.

Outside of ecological fiction, Garforth (2005, 2018) outlines a wider ecopolitical philosophy which

> builds its vision of a greener future on a desire to fundamentally transform present human relationships with nonhuman nature. Rather than seeking simply to modify global capitalism to attend to issues of resource depletion and pollution, radical ecological approaches insist that the environmental crisis demands a philosophical and cultural orientation to the natural world that would embed human societies in a much closer and less instrumental relationship to the ecosystems that support them.
>
> (Garforth, 2005, p. 2)

The environmental crisis is then a disaster but also an opportunity to rebuild the world better, to re-orientate ourselves into a lifestyle that respects the natural world. Similarly, McNeish (2017) situates mainstream environmentalism as equivalent to the amillennial theologies (Chapter 3) but also identifies radical environmental groups that have more affinity with post-millennialism, in that for them "the apocalypse is a revelation about the inherently destructive nature of the capitalist economy that can be used to inform an ecological revolution which links social and environmental justice to sustainability" (p. 1047). But the environmental utopianism can also have a more forceful radical streak to it, one that not merely seeks to use the environmental crisis as an opportunity to better society, but actively seeks it out, a tendency McNeish sees in some deep ecology groups where "there is even a sense that the prospect of an environmental apocalypse is to be welcomed as an unpleasant but necessary cleansing path to ecotopia" (p. 1047).

Conclusion: utopia, identity and the social imagination

Like almost all of the central concepts discussed in this book, utopia is a rather vague concept. I have tried here to give a brief overview of some of the main discussions in utopia studies that I think are relevant, but I have not tried to give any clear definition of what utopia is, other than the vague notion of a better society – usually, but not always, like the millennium, in the future. For Bloch (1986), the concept is very broad, encompassing daydreams and wishful thinking as well as the millennial biblical tradition and the early modern literary genre of More and his imitators. For others, utopia is strictly limited to a set of characteristics of More and similar writers, in that it needs for example to include a clear blueprint on how the society will look. With such varying definitions, Marxism is therefore sometimes included as utopian, and sometimes not, and even Popper and the cold-war anti-blueprint philosophers, as I have argued, can be seen to be utopian in the sense of wanting to theorise how society could be run better in some way.

The precise meaning of utopia, however, is somewhat secondary for me here, and the inclusion of a discussion on utopia in this book relates to the fact that it evokes a millenarian tradition (one of the various influences on utopia that

Kumar, 1991, discusses). As with other apocalyptic visions and narrative tropes, the millennium can be analysed as a continuing chain of social representations, a persistent myth in Blumenberg's (1979) sense. Second, as utopia is a specifically social imaginary, we can also try to ask further questions as to whose utopia it is, and related to that, who is not part of the utopia. This will then enable us to slot the utopia/millennium into our wider look at the apocalypse and theorise how it lets us understand contemporary existential risks.

Confronted with new concepts or information, social groups collectively make sense of them through the anchoring to already familiar concepts. This is a symbolic and social process rather than an individual cognitive one, because as Jovchelovitch (2007) notes, "the human child cannot produce herself as a feeling and thinking self without the participation of other human beings and it is symbolic because it uses arbitrary tokens to invest the object-world with meaning" (p. 3).

As with Bloch (1986), who argued for a wider definition of utopia, the hope for a better future is in some ways a universal and perfectly understandable feeling (Jovchelovitch and Hawlina, 2018) and a vision of a better future – whether as daydreams, fantasies of abundance, or political visions of how a society could be better organised. This clearly also applies to the hopes of national liberation in the nativist millennialisms (Rosenfeld, 2011) that were precursors and progenitors of the biblical apocalyptic.

Taking specific manifestations of these, however, we can follow how at least some of the more formal ideas of utopia developed as authors took inspiration from previous work or cultural traditions. Thus, as Kumar argues, the Christian millennial tradition was one of the influences of the (early modern) utopia as a literary genre, which in turn also influenced the literary genres of the Robinsonnade, the novel as a literary form, and science fiction and fantasy as particular genres within that literary form, which in turn influenced the development of utopias and dystopias in newer media such as cinema, television and video gaming. The political philosophy of Plato was of course also a direct influence on More, which he notes explicitly; this influence also coloured the humanistic and political utopias, which influenced Marxism and so on. Tracing influences and heritages behind contemporary utopias and utopia-like fantasies allows us to introduce a few observations about how a social representations approach can shed some lights on the role utopian visions play in contemporary, secular, apocalyptic fears.

In considering how a particular utopian vision gets incorporated and transmitted through various cultural and social groups, it is important to keep in mind that a utopian vision is generally a utopia not just *about* a particular society, but *for* a particular society as well. This is particularly obvious in the case of the explicitly political utopias of Plato, More and Marx, where the ideas they developed can be seen as potentially to be implemented in the society they are hoping to improve. This is also the case even in the satirical or more allegorical utopias (such as the imaginary societies in Swift's *Gulliver's Travels*) that may not develop explicit ideas the writer's society should follow, but instead offer more circumspect ways in which that society can be enlightened. In these cases, the beneficiaries of the utopia are fairly clear and delineated, even if this is rarely made explicitly so. However, we can also look at the utopia, and/or their intended readership who might take the utopia's lessons to heart and help build a better society, and ask who is not part of the scheme.

This is maybe particularly clear in utopias written in colonising and slave-owning societies such early modern Western Europe, where a whole class of people are simply written out of the picture. Plato's utopia is a utopia of the ruling classes; the early modern and enlightenment utopias primarily written in Britain and France were authored within the context of the developing colonial empires of these countries and their participation in the slave trade. It has been noted often enough that Robinson Crusoe, in so far as it can be considered a proper utopia (James, 1993), includes the "primitive" native who is rescued and then made into a man-servant by the builder of the utopia, and while the servant Friday partici-pates in the utopia, it is also clear that it is not *his* utopia. Likewise, utopias written from a male perspective can also easily become explicitly male utopian fantasies. Neville's *Isle of Pines* (Bruce, 1999), for example, includes the sexual fantasy of a sailor marooned on an island with four women (one of which is an African slave), leading to a rather patriarchal utopia, a situation noted for example and even inverted in explicitly feminist utopias such as Gilman's *Herland* (Peyser, 1992; Relf, 1993). In this sense, utopia can easily become the utopia only for some people, and not necessarily for others: Either inequality is built into the sys-tem from the start, or other societies are not being considered at all. The fact that a utopia is only a utopia for some societies or sections of societies is a recurring theme in the imagination of dystopian visions arising from utopian ideals *gone wrong*; the dystopian fantasy then merely makes explicit that utopias are limited to the value systems, ideals and standards of those who dreamed it up and recount the utopian experiment from the point of view of those it excludes.

On another level, however, utopias are tightly bound up with social group iden-tities through the group background knowledge, norms and values that bind them together. In that sense the fact that a utopia is only a utopia for that group is not so much because other groups are excluded but because the background knowledge norms and values constrain what is seen as desirable and what isn't. As Mannheim has noted, at least for some interpretations of the idea of utopia, the concept is tightly bound together with ideology, the main differences being, for Mannheim, that ideology is looking towards the past and instituting the prevalent worldview of the present, whereas utopia is looking towards the future – and as Levitas and others note, this distinction can at the end be rather arbitrary (e.g. through a utopia that is oriented towards implementing past ideologies in the future, Mannheim's "conservative utopia").

The religious utopias instituted for example in the communal living experi-ments of millennial sects (Cohn, 1957) are clearly communal and community-based projects: They may well be open to anyone willing to join, but potential membership relies on the recruit sharing the group's common values and beliefs. Religion, understood as an ideology that guides how concepts are being seen and understood, thus plays a role in the building of utopian ideals, not merely through the stipulation of a better world during the millennium (and after) but also through the shared understanding of the norms and values that will bind that future society together. The unrealised early modern utopias of Bacon and others likewise feature societies that are built on some of the prevailing social norms that define the societies from which they derive. They are often not completely new imaginaries of how a society could look but extrapolations of how society would look if it was really built around the social norms that are already accepted in real society; utopias written from a Protestant perspective, for example, feature piety, frugality and asceticism as the principles that bind.

In Chapter 5 I looked at how the social knowledge of a group is bound through the cultural dynamics of how information is processed and linked to the group identity; information that conforms with defining group norms and values and defining background knowledge gets assimilated easily while information that doesn't will get either ignored, or if compelling, incorporated through the alteration of the protective belt of norms and values that are less defining of the group. If we combine this perspective with the anthropological issues noted by Douglas (Chapter 6), we can also link the specifics of the cultural issues that clash with a group's norms and values through ritualised purity and dirt taboos. Utopias (and of course dystopias) then develop with respect to already established cultural taboos, in particular the dirt and purity laws focussed on bodily boundaries.

Chapter 8

Nuclear apocalypse and the nature of evil

Introduction: imagining the bomb

Marking the occasion of the 75th anniversary of the bombing of Hiroshima in August 2020, the UN Secretary-General António Guterres argued that it is time for nuclear weapons to be abandoned (Guterres, 2020). Having been written within a backdrop of a world in the middle of a global pandemic, the urgency of the message may have been rather lost, however even though the cold war is over, the threat of nuclear weapons has steadily been creeping (back) up the global agenda over the past two decades. New states have acquired nuclear weapons; new conflicts have emerged and old ones have heated up. A few years earlier, the former US military analyst and whistle-blower Daniel Ellsberg (2017) published his best-selling insider account of cold war decision-making outlining often hair-raising accounts of how close we sometimes came to disaster. Thus although nuclear weapons apocalyptic discourse was shaped and given form during the cold war era, it is still a concern that is very much with us, even if it is temporarily crowded out by other concerns.

From the outset, discourses around nuclear weapons have made explicit use of apocalyptic language. This is of course not entirely surprising, since the destructive capabilities of nuclear weapons dwarf that of any weapon that was previously available. Nuclear warfare relates often quite naturally to several apocalyptic tropes that I have discussed in previous chapters. As Jaspers (1961) discussed, for example, nuclear weapons can be easily interpreted as a modern-day Prometheus story, a variant of the Frankenstein myth on the dangers of knowledge, where in this case the dangerous knowledge easily evokes the image of the fire Prometheus stole. As both a literal inferno and a weapon of war, nuclear weapons equally evoke scriptural passages, mostly from Revelation, about how the end will be brought about by brimstone, fire and war. One of the four horsemen of the apocalypse is popularly interpreted as an anthropomorphisation of war itself.

With these and a host of other associations, nuclear war and nuclear weapons have shown a clear and natural affinity to a lot of religious apocalyptic, and specifically biblical, thinking. However, nuclear weapons are also a distinctly secular affair – born out of scientific research that especially during the early and mid-20th century had increasingly become distanced from religion. Unlike the apocalypses of both the religious and political ideologies, a nuclear apocalypse became a possible near-future event for people who did not hold either religious beliefs or strong political ones. In this respect, then, nuclear war was something that everybody could worry about.

The secular nature of nuclear weapons posed an interesting question for theologists who saw in it a possible way for the events of Revelation to play themselves out. As the theologian Jim Garrison (1982) argued, theology (and specifically, theodicy) needed a rethink after Hiroshima: "In order to fully appreciate the Hiroshima event", he argues, we must recognise that

> with the power of mass destruction in our hands, we have taken upon ourselves that last category attributed to God in the traditional view: the determination of apocalyptic judgement. What the apocalyptists believed was fixed by the counsel of God and brought to pass by divine will and action alone is now something within the realm of human decision.
>
> (p. 5)

Similar thoughts were also articulated by Jaspers (1961, p. 21).

Secularisation, to the extent that it can really be said to have happened at all, meant that the wrath of God, or fate, needed to have a replacement in the actions of humanity. Disasters cannot just happen, they need to be the fault of someone, and with the realisation that humanity is responsible for many evils previously assigned to supernatural forces, theology needed to adapt – this appears to be particularly so with nuclear weapons.

With the bomb representing a new way for humanity to destroy itself in large numbers, it has become an emblem of a new phase in history when destruction on a wider scale has suddenly become not just a theological possibility. However, while the destructive potential of nuclear weapons is in some way new, the suffering and death it has caused so far are much less than the atrocities that the world had stumbled out of at the time they became widespread news: The Holocaust, the genocides and the massive death and destruction that "conventional" war has caused through fire-bombing or trench warfare still manages to dwarf the death and suffering caused by the bomb at the end of the Second World War. What is it, then, that fires the apocalyptic imagination about nuclear weapons that transcends the justifiable fears over conventional wars, and conventional authoritarian and genocidal autocracies?

In some way, we can say that nuclear weapons have become an easily visualisable representation of the wider horrors of war, that nuclear weapons are symbolising war rather than having replaced it as an object of worry. Nevertheless, the weapons themselves have several qualities that make them particularly fearsome. One would be their unpredictability, the fact (or at least the fear) that, if a nuclear war breaks out, the first we will know about it is through the immediate destruction of our lives. At least if a conventional war breaks out, we would have some time to adjust and get used to the idea that our lives will change for the worse. Another thing that distinguishes nuclear weapons as particularly dreadful is the insidiousness of the nuclear fallout, the fact that even years after the initial exposure we can still fall victim to the diseases and cancers caused by radiation. The sneaky, silent killer effect of nuclear fallout is certainly unsettling, evoking a similar dread as cancer does in general with respect to other similarly (or even more) deadly diseases such as heart disease. It is also an instance of what risk psychology researchers (e.g. Slovic, 1987) have called a "dread risk", that is a low-probability but high-impact risk, a combination that, similar to terrorism, scares us more than equivalent risks with higher probability but less of an impact (Gigerenzer, 2004). This somewhat unique combination of dread associated with nuclear weapons was analysed by Michael Ortiz Hill (1994), who explored the apocalyptic nuclear dreams of a collection of 100 people, finding it a "shadow bomb" with an "ambiguous nature as an object of *holy* terror" (p. 1, original italics).

This chapter will attempt to look at the central question of the *meaning* of the bomb with respect to the apocalypse and the ambiguous position it holds between the secular-scientific and the prophetic-religious discourses it spawned, the fears and the hopes – but mainly fears – it engendered and what the legacy of the cold war era atomic bomb discourse is. Now, 30 years after the end of the cold war, the bomb has made itself relevant again through the acquisition of nuclear weapons by a whole new set of nations (India, Pakistan, North Korea), but without the comforting ideological and religious certainty of the cold war.

The following section will introduce very briefly the history of the making of the bomb and its role in cold war Western society (concentrating mostly on US, UK and West German contributions), though there are also interesting histories to be told from the other side of the cold war (Dobson, 2016, 2018). I will then look at some of the main themes from the previous chapters and apply them to nuclear discourses, focussing in particular on the eschatological role of (secular) evil and how we can think about our social representations of the bomb as an interplay of the secular-religious apocalyptic traditions.

Making the bomb

From the outset, as physicists realised the enormous amount of energy that could be released through nuclear fission, thoughts and imagery turned to apocalyptic scenarios. As early as 1904, the Cambridge physicist Frederick Soddy wrote that "the man who put his hand on the lever by which a parsimonious nature regulates so jealously the output of this store of energy would possess a weapon by which he could destroy the earth if he chose" (quoted by DeGroot, 2004, p. 5).

Some of this realisation was famously formulated in apocalyptic science fiction, with H. G. Wells' scarily prophetic 1914 book *The World Set Free* (Wells, 1914) describing an atomic world war that destroyed cities and ended with the technology being used to provide peaceful energy. The novel also contained the millennial/utopian speculation that a technology this destructive would also make war unthinkable and usher in an era of peace and plenty, foreshadowing the "mutually assured destruction" doctrine of the cold war.

By the 1930s, several physicists such as Leo Szilard, Ida Noddack and Enrico Fermi started looking at the chain reactions that could be achieved if certain atoms were bombarded with particles, with Szilard taking particular inspiration from the bombs described by Wells (DeGroot, 2004, p. 17). Being instrumental in the early stages of developing the mechanisms for a chain reaction in 1939, Szilard also tried to warn about the destructive potential of the bomb (Weart, 1988, p. 79).

The destructive potential of any bomb was realised in Germany as well, which also occasioned "conflicts between duty and morality" (DeGroot, 2004, p. 21) by German scientists, who were tasked with researching building the bomb. Some of them, like Werner Heisenberg, claimed after the war that they only pretended to collaborate on this, and without great enthusiasm – whether this is believable or not, at the time it appeared that the prospect of a German bomb was "frighteningly real" (p. 20). Partly through the fear of a German bomb, American and European émigré scientists, among them those who had large reservations about developing the bomb, such as Einstein and Szilard, convinced President Roosevelt to start their own bomb research.

The story of how the Manhattan Project was born and developed has been retold enough elsewhere (Hughes, 2003; DeGroot, 2004; Thorpe, 2008), but what is fascinating in these accounts is the amount of apocalyptic fear and expectation that was present at Los Alamos. Also interesting is how, as Thorpe recounts, these fears and apprehensions on behalf of the scientists working there were subdued through concentration on the day-to-day work, through gallows humour, through the thorough militarisation of the command structures, or through the simple thought that they were racing the Germans and therefore the Americans having the bomb first being the lesser of two evils.

The apocalyptic apprehension, however, was always close to the surface. J. Robert Oppenheimer, the director of the lab and later the publicly acclaimed "father of the bomb", was himself deeply ambivalent and conflicted about it (Thorpe, 2008).

After the first nuclear bomb was exploded on the Trinity test site in New Mexico in July 16, 1945, Oppenheimer, a well-read man who used some of his spare time learning Sanskrit, famously quoted a line from the *Bhagavad Gita*:

> If the radiance from a thousand suns
> Were burnt into the sky,
> That would be like
> The splendour of the Mighty One –
> I am become death, the shatterer of worlds.
>   (DeGroot, 2004, pp. 64–65)

This famous remark shows how easily it came to the scientists working on the Manhattan Project to link the momentousness of the events they were part of and responsible for to this prophetic and religious language as a way of expressing their underlying fears. It is quoted in almost every account of the history of the Manhattan Project, indicating that Oppenheimer appears to have captured an evocative and widely shared sentiment. While in this case this is language rather than religious belief – there was no indication that the *Bhagavat Gita* was for Oppenheimer anything more than an interesting text from a culture whose language he was studying – the associations between the bomb and the mystic, religious and supernatural was clear in other remarks as well. And this is true not just of Oppenheimer and his scientists, but also of those who disapproved or had misgivings about the bomb. President Truman himself, upon reading (the project's military director) General Lesley Groves' report of the Trinity test, commented: "We have discovered the most terrible bomb in the history of the world. . . . It may be the fire destruction prophesised in the Euphrates River Era, after Noah and his famous Ark" (quoted in DeGroot, p. 80). This recurring mythic register of speech was called the "atomic sublime" by Hales (1991). The mythic and religious language was pervasive, as outlined by Weart (1988, ch. 5); the very name chosen for the Trinity testing site alludes to the foreboding, mystic and almost religious experience that many people recounted as experiencing.

The frequent use of apocalyptic language did not mean the people involved (politicians, military and scientists) were against the deployment of the weapon, but this shows that attitudes were far more ambiguous and ambivalent. Despite the quote above, Truman himself was of course ultimately responsible for Hiroshima and Nagasaki, and like him many others persuaded themselves that the use of the weapons ultimately saved more lives through shortening the war than they took. This has been debated, not just with hindsight but at the time, at length.

And the bomb was an ambivalent beast in many senses. As Boyer (1994) described in detail, the mood in the US after Hiroshima ranged from horror to acceptance of the bomb as a lesser evil to triumphalism. The bomb heralded a new era of technological optimism, with the word "atomic" serving as a marketing device to sell vacuum cleaners and cocktails (Wojcik, 1997, p. 102). In the 1950s people picnicked near the Nevada testing sites, watching the bombs go off, spawning a minor tourism boom in Clark County (Boyer, 1994). Lingering health hazards of radiation poisoning were at first not widely known or appreciated, and they were hushed up by the military.

The full horror of the bomb did come to prominent public light, however, particularly so through a widely read and influential article by John Hersey (1946) in the *New Yorker* magazine. This article, immediately re-published as a book, detailed for the first time eyewitness accounts from Japanese survivors. This personalisation of the previously more abstract, imagined suffering occasioned widespread introspection about the nature of atomic warfare – but wider, apocalyptic anxieties were never far from the surface. As Wojcik argues, the bomb symbolised (among other things of course) the automated culture of the new mid-20th-century technological era: "The push-button efficiency of dishwashers, television sets, washing machines, and vacuum cleaners simplified life and provided entertainment, but the same technological efficiency had made a push-button apocalypse a reality" (Wojcik, 1997, p. 103). The famous "red button", or "doomsday button", has come to symbolise another one of the specific "dread" elements that distinguishes the bomb from the other existing ways of killing huge amounts of people. The convenience of a single button that decides the fate of the earth and rests with the whims of just one person is something that is difficult to contemplate, particularly so when the responsibility for the button rests with people who are generally held to be less psychologically stable. This is a fear that is still very much with us, 30 years after the end of the cold war.

Many scientists from the Manhattan Project kept their enthusiasm about the technology they helped birth, and its purported propensity to keep the peace – notably so Edward Teller, who would subsequently become help develop and champion the even more powerful H-bomb. Others were troubled, not just by the actual harm and suffering the bomb had caused in Japan, but also about its pro-pensity to bring the world to an end. Most visible among these was Oppenheimer himself, who used his newfound national prominence as the "father of the bomb" to attempt to start national conversations around nuclear weapons. This ultimately earned him an appearance before the House Un-American Activities Committee (HUAC) as part of the McCarthy hunt for communists, the revocation of his secu-rity clearance and his ultimate retreat from public life (Thorpe, 2008). As Thorpe outlines, Oppenheimer's fall from public grace heralded a larger era in which scientists withdrew from participation in politics and into a purely intellectual life, which propagated a more widely perceived attitude still quite evident today that science is (or should be) merely concerned with discovering truths from nature, and that scientists should stay out of moral and political debates surrounding their discoveries.

While apolitical science may have become a norm during the cold war era, there were still very prominent scientists who rebelled against this and tried to draw attention to the role that science and technology can have in the potential destruc-tion of the world. Growing out of a groups of disillusioned Los Alamos scientists, like Oppenheimer, the Bulletin of the Atomic Scientists was founded in December 1945 by the physicists Eugene Rabinow and Hyman Goldsmith, together with the social scientist Edward Shils, to become a "moral forum for the atomic commu-nity" (DeGroot, p. 116). The Bulletin first included the famous "doomsday clock" on the cover of the June 1947 issue, setting us at seven minutes to midnight as a metaphor for how close we are to the end of the world. It has been updated regu-larly ever since, moving closer or further away from midnight depending on their assessment of the geopolitical situation. Having recently also started to include considerations of climate change, the clock, set on January 23, 2020, stands as close as 100 seconds to doomsday – the closest to midnight it has ever been (Bul-letin of the Atomic Scientists, 2020).

The history of the cold war can be reconstructed just by looking at the time-line of the doomsday clock. In 1949, it was advanced to 3 minutes to midnight in response to the acquisition of nuclear weapons by the Soviet Union. In 1952, in response to the development of the even more devastating thermonuclear H-bombs, the clock was advanced to 2 minutes to midnight. In 1960, it was moved back to 7 minutes to midnight in response to a partial easing of international tensions. It moved back and forth between 7 and 12 minutes to midnight over the next few decades, then to 4 minutes in 1981 and 3 minutes in 1984 (during the presidency of Ronald Reagan). It was then slowly dialled back, reaching 17 minutes in 1991 at the end of the cold war – the furthest it had ever been from midnight. The dial then gradually crept back up over the last 30 years, representing the gradual realisation that the end of the cold war did not really mean the end of international conflict. Climate change and environmental deterioration were also taken into consideration.

The doomsday clock was (and still is) a devastatingly simple rhetorical device to distil the perceived danger of nuclear weapons into one easily communicated number, and although the time itself is set through qualitative political judgement, the fact that this judgement is represented through a number by a society founded by atomic scientists gives it a perceived preciseness that it doesn't of course quite deserve. The metaphor of the clock was also an inspired choice, the clock representing the inevitability and the urgency of the issue. Horn (2014, p. 80) sees the clock's significance in symbolising that with the invention of nuclear bombs an irreversible step had been taken: we can disarm and we can find political solutions international cooperation that makes their deployment unlikely, but we cannot uninvent them.

Meanwhile in Britain, the UK joining the nuclear powers in 1952. A pacifist countermovement, led by a collection of scientists, philosophers, writers and politicians (prominently among them the enormously influential philosopher and polymath Bertrand Russell, the novelist J. B. Priestly and the physicist Joseph Rotblat), established the Campaign for Nuclear Disarmament (CND) in February 1958. Originally intended to be a "high-level pressure group" similar to the Bulletin, the CND would go on to develop a broad public appeal that would "give focus to [ordinary peoples'] fears about nuclear weapons" (DeGroot, p. 230). The CND transformed itself into a mass movement largely affiliated with the Labour Party, even while, as DeGroot notes, popular support for the unilateral disarmament demanded by the CND never reached more than 33% (p. 232). The CND diversified itself with a move into more general environmental activism after the 1960s (Burkett, 2012) as a part of a wider contemporary shift from anthropocentric to ecocentric views.

While the CND lacked the neat apocalyptic metaphor of the Bulletin's doomsday clock, the apocalyptic urgency through which it and its supporters described their mission was clear. Bertrand Russell for example, writing in 1959 in an essay on how the situation could be diffused internationally, argued that

> our planet cannot persist on its present courses. There may be war, as a result of which all or nearly all will perish. If there is not war, there may be assaults on heavenly bodies, and it may well happen that means will be found to

cause them to disintegrate. The moon may split and crumble and melt. Poisonous fragments may fall on Moscow and Washington or on more innocent regions. Hate and destructiveness, having become cosmic, will spread madness beyond its present terrestrial confines.

(Russell, 2010 [1959], p. 5)

The almost biblical imagery of destruction that Russell conjures up here, complete with the "falling star" motif (Revelation 9:1, KJV) of a crumbling moon is rather surprising coming from a philosopher known for his level-headed agnosticism that regards religion as mostly superstition (Russell, 1957). But this demonstrates part of the mood of the time, when science showed the push-button ease with which ever more powerful and destructive weapons could be unleashed at the whim of just one man, and with the atrocities of the war the world had only recently stumbled out of showing the madness collective humanity is capable of.

The mood was maybe even more collectively sombre and subdued in Germany. After the immediate post-war years and the rise of a new generation, the country gradually if not always successfully attempted to come to terms with its guilt; and the destruction and atrocities that humans were capable of were felt more keenly in Germany because they couldn't shift the blame for the war onto the other. As a consequence, Germany (along with Japan, unsurprisingly) developed a form of nuclear pacifism and fatalism. Split in the middle by the Iron Curtain, with its respective allies' nuclear bombs stationed on its territory, the fear was that Germany would become the primary battlefield if the cold war ever escalated into a nuclear confrontation (Ziemann, 2016; Sonne, 2018).

Not being a nuclear power itself (and forbidden to be so by the 1954 treaty of Brussels), Germany did not have a clearly focussed anti-nuclear lobby in the form of the Bulletin or the CND. Nevertheless, anti-nuclear voices were probably the more dominant in Germany compared to the comparatively still enthusiastic public support in the US, UK (and France later on, which joined as the next member of the nuclear club in 1960).

As in the US and UK, atomic scientists themselves participated in trying to sound the apocalyptic alarm over nuclear weapons in Germany, though the country's recent history – and their involvement in it – made some of their reflections rather awkward. Some of the most prominent scientific advocates of nuclear disarmament were scientists like Carl Friedrich von Weizsäcker (Neuneck, 2014), Otto Hahn and other members of the group of atomic scientists who worked under Heisenberg on the Nazi bomb project. Eighteen prominent German nuclear scientists (including many members of the Nazi project), participated in 1957 in drawing up the "Göttingen Manifesto", declaring opposition to then Chancellor Adenauer's plans to arm the Bundeswehr with tactical nuclear weapons. Similar to Russell's worry about the potential future development of even more destructive weapons than the H-bomb, the manifesto projects a possible future where "no natural limit is known for the developmental possibility of the life-destructing effect of strategic nuclear weapons" (Universität Göttingen, n.d.), otherwise however being somewhat less apocalyptic (imagining the destruction *merely* of Germany rather than the whole world).

Overall, though, given their involvement in the Nazi bomb project, von Weizsäcker's (and his colleagues') impact and legacy were fiercely fought over and clouded their anti-nuclear engagement in later life (Schäfer, 2013). Protestations were made that they had no choice to participate and that their progress was deliberately slow. The extent to which these protests were true and how much was an attempt at saving face is difficult to judge, but von Weizsäcker at least attempted to draw a line under his earlier life to become a prominent campaigner for disarmament.

The German/Swiss existentialist philosopher Karl Jaspers wrote theologically and humanistically more reflectively on the bomb in his post-war work. The fascist and communist ideological regimes of the 20th century were "the result of a decline in political humanity and of an increasing primacy of modes of technical or instrumental rationality, which erode the authentic resources of human life" (Thornhill and Miron, 2020). Believing it to be a result of oppressive government, he opposed technocracy and believed that "German society was not sufficiently evolved to support a democracy, and Germans required education and guidance for democracy to take hold" (Thornhill and Miron, 2020). In fact, he regarded totalitarian government as the only important problem that is equivalent to the atom bomb (Jaspers, 1961, p. 15). Regarding specifically the bomb and the language he used to discuss it, Jasper's own experience of the Nazi era maybe influenced a mood that was similarly negative to the apocalyptic imaginaries discussed above: "Experts are saying with complete certainty, that through the actions of humanity it is today possible to bring about the total destruction of life on earth" (1961, p. 13, my translation). His critique of totalitarian government was influenced by the cold war ideology of mutually assured destruction, that a war of the superpowers would "encompass the probability of the demise of humankind" (p. 123, my translation). But this alone did not reassure him that neither superpower will never use it.

As the doomsday clock shows, fears over nuclear war have been fluctuating up and down since the cold war. But nuclear fear has never really gone away, even after the collapse of the Soviet Union. Jonathan Schell's (1982) influential book on the possible effects of nuclear war, published in the early years of the Reagan administration, envisages a "republic of insects and grass" as his apocalyptic vision for a post-nuclear war US. As an illustration of the impact of Schell's vision, in my own second-hand copy of Schell's book, the previous owner had scribbled on the margins that "we should educate our children (all the world) about nuclear holocaust from grade 1. By the adolescent years, they are Rambo-ized".

Nuclear fears were also intimately intertwined with arguments surrounding nuclear power: The incidents in Three Mile Island in 1979, Chernobyl in 1986 and Fukushima in 2011 cemented in public imagination the link between nuclear war and nuclear technology in general, with Chernobyl in particular offering a ready-made picture of what nuclear devastation looks like. New nations such as China,

Israel, India, Pakistan and North Korea have joined the nuclear club, and insider accounts such as that of Ellsberg (2017) ensure that nuclear war remains part of a wider apocalyptic possibility, even in times where our eschatological fears are dominated by climate change and pandemics.

Apocalyptic themes

The previous section primarily gave a brief flavour of the apocalyptic language in which the atom bomb has been described rather than a clear history of the bomb. Other discourses were, of course, also available (Weart, 1988), so these were not the only ways through which the bomb had been made sense of. This section will briefly outline how some of the nuclear discourse relates to the sociological and theoretical issues I have discussed in previous chapters.

Religion and science

The influence of religious imagery and thought on atomic discourse has been large and pervasive. As I tried to show above, religiously influenced discourse was memorably demonstrated by prominent humanists like Russell and scientists like Oppenheimer, indicating the influence religion and religious language and metaphor have had on how the bomb is being processed.

None of this is particularly new, as Weart (1988) documented; the religious influence on atomic discourse has been substantial. This extends even to the atomic discourses that happened within the nominally atheist Soviet Union (Dobson, 2018). But the influence of religion on thought about the nuclear bomb and the cold war extended beyond a general mystic or doom-laden discourse of Russell or Oppenheimer that borrowed religious imagery, intentionally or not. Religious communities that had already been interested in the end times through their millennial doctrine quickly understood the apocalyptic potential of the bomb to be part of the signs and omens that portended the approaching cataclysm. Nuclear weapons were the (or at least one of) the means through which the events of Revelation would play out.

This was of course most influentially the case in the US, where evangelical premillennial and dispensational denominations had been developing in influence since the Millerite prophecy, and which after the war developed even further into a formidable challenge to the mainstream Protestant and Catholic denominations that. This is still very much ongoing, perhaps with less emphasis on nuclear weapons. Biblical language in Revelation and other places had frequently highlighted fire, brimstone and similar methods of destructions as part of what awaits us in the apocalyptic future, and the parallels with the effects of the atom bomb are easily drawn – not just in the mystic, metaphorical sense discussed above but within the Protestant tradition of reading the Bible literally.

In *The Late Great Planet Earth*, one of the best-selling non-fiction books of the 1970s, Hal Lindsey (1970) constructs a larger argument on how biblical texts shows us that the end of times is near. For example he makes the following connection, discussing a passage in Revelation on an invading army from the east:

> A terrifying prophecy is made about the destiny of this Asian Horde. They will wipe out a third of earth's population (Revelation 9:18). The phenomena by which this destruction of life will take place is given: it will be by fire, smoke (or air pollution), and brimstone (or melted earth). The thought may have occurred to you that this is strikingly similar to the phenomena associated with thermonuclear warfare.
>
> (1970, p. 82)

While Lindsey was one of the most prominent evangelical voices working out this new nuclear theology during the latter stages of the cold war, these re-interpretations of the nuclear bomb as the prophesised "fire and brimstone" became part of the evangelical mainstream just as the evangelical denominations themselves moved further into the national mainstream.

Lindsey and fellow premillennial prophecy interpreters wove the nuclear bomb and the wider cold war context into a larger apocalyptic framework that incorporates the many often cryptic biblical passages of Revelation (and a few select other apocalyptic passages) into a fairly coherent interpretation of contemporary political events (see Lahr, 2007; Boyer, 1994 for more detail and discussion on the wider social context of this movement and the internal differences within it). In very broad strokes, with the main opponent of the cold war being an officially atheist country, the conflict is easily cast in theological terms as a struggle between good and evil.

The post-war flashpoints in the Middle East that came with the establishment of Israel and its subsequent struggles with its neighbours also helped cast the conflict into biblical terms as a precursor to the struggles prophesised in Revelation to occur in the Holy Land: "Since the restoration of Israel in 1948, we have lived in the most significant period of prophetic history" (Lindsey, 1970, p. 62). However, further prophecies need to be fulfilled in order for the events of Revelation to take place. One of them is that the third temple needs to be rebuilt, and for that to happen the Dome of the Rock stands in the way: "Obstacle or no obstacle, it is certain that the temple will be rebuilt. Prophecy demands it" (Lindsey, 1970, p. 56).

The role of Russia is that of the biblical hordes of Magog, the invader from the north, owning to the curious coincidence that Moscow is almost exactly due north from Jerusalem, and that it sounds slightly similar to Mesech, and Russia generally being similar to Rosh, both mentioned as princes of Magog in Ezekiel 38 (Lindsey, 1970, p. 65). Various other bits of prophecy are found in contemporary politics as well, with the Treaty of Rome that established the European community, for example, being the ten-nation coalition led by the Antichrist from Rome.

Cold war biblical prophecy extends beyond the bomb focus of this chapter, yet these examples do highlight the religious/secular ambivalence of the atomic bomb and how seamlessly it can move between communities and find new meaning relevant to them. As Lahr (2007) argues in a summary of a 1960 essay by Perry Miller:

> The difference between secular and prophetic forecasts of atomic destruction hinged on a framework of meaning. While prophecy writers who spoke and wrote in the beginning of the atomic age expounded on a future that had theological significance, secular Americans wrote of a meaningless future. For evangelical premillennialists, history as fulfilled prophecy proved a divine order and justified their belief system. (p. 18). . . . [I]n fact, the nuclear age brought these two strands of apocalypticism together.
>
> (p. 26)

The bomb and apocalyptic culture

The often repeated contention that we live in a largely secular age, and the fact that the bomb is an unmistakably secular, scientific achievement, has led some commentators to distinguish it as a secular rather than religious apocalypse. Wojcik's (1997) analysis of the atom bomb within the context of US millenarian culture took this view, despite the fact that the bomb was an unmistakably large part of the cold war evangelical mythology. The distinction between secular and religious, as I hope to have shown in Chapter 2, is not quite as straightforward. Writers such as Jaspers and Garrison have noted that even mainstream theologies need to find a place for an interpretation of what the atom bomb means. Even without a direct, literal re-interpretation of Revelation in current affairs, the bomb, along with the wider aftermath of the Second World War, has opened up often uncomfortable questions about the nature of evil and humanity's destructive potential.

The "atomic sublime" that Hales (1991) noted, and the often straightforward connotation of the fiery atomic destruction with passages from the Bible as well as other religious texts from quite different cultures, as well as the theological/scientific ambivalence that Weart (1988) wrote about, eased its incorporation into the apocalyptic imagination much better than some of the other horrifying man-made atrocities that the 20th century gave us. The catastrophic suddenness, the fire and brimstone, the immediate nuclear winter, the capriciousness and unpredictability of "the red button" and the ultra-fine balance of peace ensured only through mutually assured destruction gave the bomb a cataclysmic, apocalyptic quality that more slow-burning but just as destructive catastrophes like climate change simply lack. This has then served as an example for how people who are worried about climate change intend to portray it as a more urgent issue.

The world imagined through the atom bomb becomes one where our fears and hopes about prophecy can be re-evaluated from both a religious and a secular starting point where, as Lahr (2007) argued, the secular and the religious strands of apocalypticism can come together. Like risk discussed in Chapter 6, the bomb can be argued to form a boundary object in the sense of Star and Griesemer (1989), whose interpretative flexibility enables different groups to communicate through a shared representation that is anchored to subtly different background knowledges and values. In that vein we can also look at the bomb as a social representation in Moscovici's sense (Chapter 4), which we interpret and make sense of through our own preconceived knowledges and myths. Weart (1988) for example notes the frequent allusions to the Prometheus and Frankenstein myths that have shaped our ideas about nuclear technology, particularly in its early days, the way nuclear physics was seen to have uncovered the "forbidden knowledge" that has a long history in Abrahamic traditions. And the bomb has been objectified. While already an object concrete enough in its own right, the *idea* of the bomb has itself become objectified through for example the extensive personification of the bomb as a (usually male) anthropomorphisation. This is illustrated by the "Little Boy" and "Fat Man" code names given to the Hiroshima and Nagasaki bombs, respectively; these personifications are among the enduring images and narratives that Weart has identified. Even academic analysis takes this personification forward, for example in DeGroot's title of his book alluding to it being a biography: *The Bomb: A Life.*

Thus the ambivalence of the bomb is more than just a collection of different interpretations of its meaning, instead it allows the different interpretative communities, such as the scientific and secular world and the theological, millennial world to share and inhabit the same interpretative space (similar to the unifying role UFOs have played as simultaneously scientific and esoteric objects, as argued by Robertson, 2016): It allows secular and even atheist political commentators and campaigners to find a voice and a connection through which their worries can be articulated that manages to travel outside their community.

Thus there is a real connection between the secular apocalyptic and the millennial movements that scholars in millennial studies have tried to analyse. Despite his rather strict delineation between secular and religious apocalyptic, this was also recognised by Wojcek as he did include a long chapter on nuclear and other secular apocalyptics in his book. If this connection is real, it might be worth looking at how the sociology of religion that influenced classic millennial studies literature can answer some interesting and maybe uncomfortable questions about the nuclear apocalypse. If the question that the scholars reviewed in Chapter 3 (such as Cohn, Barkun or O'Leary) wanted to answer about millennial movements was what makes them so attractive and their rhetoric so persuasive, then we might be able to use a similar analysis to ask, why is/was the nuclear apocalypse such a persuasive perspective even for secular humanists like Bertrand Russell? Plenty of voices have tried to argue that the nuclear war is neither as imminent nor, if it were to happen, as destructive as usually claimed (e.g. as recently as Mueller, 2010), however these voices remain marginal. This is not to argue about whether Mueller's case is justified or not, but it is an interesting counterpoint to climate change, where the contrary voices have had a large and enduring influence on the public conversation.

The answer to this may then at least partly be because of the special narrative connection that nuclear war has to our apocalyptic cultural heritage, which is much more lacking in climate change, and that an apocalyptic narrative has certain features which make it attractive. First, there is the contention that the apocalyptic, like religion more broadly, is of most relevance to marginalised sections of a society, as argued by Cohn's historical survey of medieval millennial movements. Applying this to the nuclear fears, we can deploy the argument fairly directly to the marginalised evangelical and dispensationalist movements that have adapted the bomb as part of their general millennial belief system. Among "secular" campaigners against the bomb, marginalisation can be seen in the left-leaning nuclear campaigners that set their worldview in opposition to the authorities, certainly in the US, that in the early phases of the cold war equated opposition to nuclear armament with communist sympathies and as such ruined the career of Oppenheimer and others. Similar marginalisation can be found in the reception of the work of the CND and the 1960 and '70s counterculture that took up anti-nuclear beliefs as part of a wider ecological worldview.

However, this link doesn't work particularly well. First, the marginalisation of nuclear protestors, to the extent that it happened, was a result of their secular millennialism rather than the cause. It would also be unrealistic to describe as marginalised people prominent scientists like Oppenheimer and Einstein or prominent philosophers like Russell and Jaspers (in Russell's case, also being a member of the UK aristocratic establishment). Similarly, if we want an explanation that specifically contrasts our responses to nuclear war with that of climate change, it's clear that marginalisation has been happening there as well. In short, a pure "relative deprivation" explanation of nuclear millennialism suffers from too many counterexamples to provide a clear explanation, echoing Barkun's (1974) general criticism of Cohn and the Christian millennialism he analysed. Marginalisation or relative deprivation might be a cause of millennialism, but it's neither a necessary nor a sufficient one.

Barkun's own elaboration of what is needed to spawn millennial movements – a disorientating societal disaster – might indeed be an improvement. Nuclear technology was born at the end of one of the most devastating worldwide crises, which compares well to the devastating crises that Barkun was thinking of. On the other hand, however, fears of nuclear Armageddon do not seem to have much diminished in the 75 decades since they were first deployed.

We can find the topoi that O'Leary wrote about – Evil, Authority and Time – in nuclear bomb discourse as well, and I believe it is in particular the topos of evil that is worth highlighting. Although it doesn't appear to have been worked into a coherent theodicy within secular and non-millenarian theology bomb narratives (unlike the evangelical interpretations of the cold war outlined above), some of the more colourful descriptions of the bomb do clearly evoke evil as an explanatory concept, with descriptions of Oppenheimer as a modern-day Dr Faust, evoking the Prometheus myth as an anchor for the bomb. Oppenheimer was described as "ambitious, individualistic, immersed in science and culture" in Chilton's analysis of media language, an image that Chilton underlines with a quote from Oppenheimer himself: "We did the Devil's work" (Chilton, 1982, p. 100). This compares also to Jaspers' (1961) allusion to the Manhattan Project scientists as modern Prometheuses who were stealing fire from the gods and earning terrible punishments. This search for evil as a way of finding meaning in the bomb shows up in our struggle to find ultimate responsibility; it shows up in the dilemma that theologians like Garrison (1982) struggled with when he wrote that Hiroshima requires a new approach to theodicy.

If Oppenheimer, Teller and the other nuclear scientists were indeed doing the work of the devil, are they then still responsible for the evil they cause? Or are they the tragic but ultimately well-meaning figures of Faust or Prometheus, or the genetic scientists of Jurassic Park? In a way, the religious "atomic sublime" discourse can help absolve the scientists and fits their creation into a wider cosmic struggle. On a secular level, this absolution can be found in arguments (such as Ord, 2020) that if not invented by them, the bomb would have been invented eventually anyway because the basic physics does not depend on Oppenheimer's work; all we need is the scientistic conviction that science will ultimately uncover all of nature's mysteries. And at least they weren't German scientists.

Then finally, in a contrasting but still secular setting that decidedly does not absolve the scientist, the evil of the bomb has been found in the secular banality of the scientist functionary who set about his terrifying work on a day-to-day basis, narrowing his horizons to the tasks ahead and leaving the decision-making – and the responsibility – to the military superiors (as described at length in Thorpe, 2008). By working on humanity's eventual extinction, the nuclear scientist becomes similar to the evil in Hannah Arendt's observations on the banality of evil in *Eichmann in Jerusalem* (1963), contributing to our new, secularised conception of what we might mean by evil when there is no God.

"Failed" nuclear prophecies and risk discourses

If we can say that the nuclear apocalypse has its prophets, evangelical, "mainstream" theological but also secular ones like Russell, then it would also have its prophecies. Confident projections into the catastrophic nuclear future have been made, and abundantly so. Admittedly they were often accompanied with qualifying phrases such as "if humanity doesn't change", though usually they are quite specific in their expectations on what is to come. The dispensationalist nuclear prophets did occasionally provide precise dates; the lessons of Millerism had come to be unlearned somewhat when they were caught up in the nuclear enthusiasm of the cold war.

But usually, prophecies of nuclear doom, particularly the more secular ones, made reference to less specified dates into the usually but not always near future, meaning that the failure of nuclear apocalypse to happen doesn't strictly refute anyone's forecast. However, when the more detailed pictures get painted, references to actors and circumstances that no longer hold, can in a sense be seen as clear refutations of the prediction. This was of course particularly so with the fall of the Soviet Union and the end of the cold war. The stalemate and enmity between the two superpowers had been a central element of many secular and religious predictions of nuclear war, so the fact that this situation had suddenly and obviously resolved itself posed a clear equivalent to failed prophecy/failed prediction.

For dispensationalist preachers and prophets like Hal Lindsey, the end of the cold war did indeed cause some initial awkwardness and reflection, but at the end the main message of the prophecy could be rescued with some fairly minor modifications to the periphery (Lindsey, 1994). And if not the Soviet Union, then at least Russia is still there, and it is still a nuclear power.

The end of the cold war also at first seemed to disprove secular prophets such as Schell (1982), and correspondingly the editors of the *Bulletin of the Atomic Scientists* set their doomsday clock the furthest from midnight it had ever been in 1991, though it is also noteworthy that even 17 minutes to midnight still conveys a note of high urgency. And over the following three decades the doomsday clock slowly towards midnight again – not merely a result of other catastrophes like climate change being added to the picture but also a result of the fact that not only do nuclear weapons still very much exist in the traditional nuclear powers, but that over this time a small trickle of nations have joined the club. One particular worry here is that of a new, smaller cold war developing between the newly nuclear nations and traditional adversaries of Pakistan and India, and the fact that North Korea, one of the most autocratic and unpredictable regimes, has developed nuclear weapons. Added to the traditional state actors to worry about, some have also started worrying about nuclear weapons falling into the hands of non-state actors such as terrorists.

These small alterations at the periphery to the overall secularised prophecy of nuclear doom have allowed refutations such as the end of the cold war to be relatively minor irritations to the big picture. The date has been recalculated and the specifics have been modified, but the overall message stays.

The risk of nuclear weapons (and nuclear disasters arising from nuclear energy) is one of the main examples for the globalised risks of late modernity that motivated Beck's (1992, 2007) risk society thesis, even though this particular threat took second place in Beck's writings to issues like climate change. However the matter of the bomb having become a "reflexive" issue through which society realises its own contribution to potential destruction has been made clear – even more so in that the bomb may have been the very first reflexively modern risk and as such precipitated an industry of theological, moral and political rethinking of the nature of evil and humanity's relation to it.

More interesting for me is Douglas' realisation that risk perception is a community, cultural affair, with the risk being introduced through the taboo-breaking behaviour of outsiders. Clearly, despite the bomb being a global concern, actual reactions to it split along community lines, with internally coherent but contradictory responses formulated for example as a catastrophic millennialism by the dispensationalist churches, and an avertive or progressive millennialism by theologians of more mainstream denominations. The precise nature of the taboo that the bomb (i.e. its makers and/or political enablers) is breaking depends on the way already established community norms and values interpret the nature of the bomb with respect to pre-established taboo – the evaluations of the nature of evil and its place in world history by theological debate, but also the secular and humanistic reflections of evil. Not all social groups participated in the reflection of how the bomb violated established group norms and values, or they did and found no clear violation – after all, the groups worried about the bomb are in opposition to groups that are not worried, or believe the bomb to either be a "necessary evil", an overblown danger or an outright good thing. Taking the analysis of reactions to the bomb from the larger sociological viewpoint of Beck to this more fine-grained community level, I believe, lets us get a clearer grip on how the bomb has shaped different ways of thinking and the interactive dynamics between the different millennial and non-millennial interpretations.

Casually reinterpreting the discussions on the "nature of evil" here with the dirt taboos as analysed by Douglas, as I have done here, is of course taking a good amount of interpretative liberty. We could make the argument that as a deadly weapon, the bomb violates the bodily boundary through several ways, and particularly so in its silent radioactivity the fallout resembles the disease pandemics that I have argued make for a good example of a Douglas-type analysis. However, any weapon designed to kill invades bodily boundaries, and while the radiation is a novel attribute of the bomb, it does not appear to feature quite as heavily in the evil discourses as its (for want of a better word) conventional destructiveness. In the end, whether my argument on the evil of the bomb holds in the strict sense that Douglas conceived of as a dirt or purity taboo or not, the bomb nevertheless appears to challenge different communities' norms and values, including their taboos, and their reaction to the bomb is inflected by how this new reality gets interpreted in the light of these values.

Chapter 9

Environmental apocalypse and the nature of nature

Introduction: apocalypse through lack of locusts

In February 2019, several newspapers were reporting on a review paper in the journal *Biological Conservation* (Sánchez-Bayo and Wyckhuys, 2019) on the loss of insect biodiversity. The paper noted in its conclusion that "the repercussions this will have for the planet's ecosystems are catastrophic to say the least" but was otherwise as temperate in its language as would be expected of a scientific paper. In the news media, the situation translated into clearly apocalyptic language. In February 2019 in the UK, the *Guardian* (a generally left-leaning and environmentally concerned newspaper) led with the headline the "collapse of nature" (Carrington, 2019a). Revisiting the issue in November of that year, the headline became "'Insect apocalypse' poses risk to all life on Earth, conservationists warn" (Carrington, 2019b).

The insect apocalypse is a potentially civilisation-ending concern not just because whole ecosystems rely on insect biodiversity but also because agriculture relies on insect pollination and therefore a large enough loss of insects will lead to catastrophic famines. The choice of apocalyptic language, both by the scientists and the media outlets who quote them, is in my view clearly appropriate.

However, the issue has also faded rather quickly from public (and media) consciousness, with the other environmental catastrophes such as climate change and (at the time of writing) the coronavirus crisis taking most of the attention. And biodiversity loss, of insects as well as other species, is only one of a host of environmental problems other than climate change that, although often interrelated, can also be seen as clearly separate apocalypses, all frequently expressed in apocalyptic language and all competing for national attention. There is industrial pollution, with attendant issues such as acid rain and the destruction of the ozone layer; there is nuclear waste and radiation pollution. Resource depletion is also an issue – maybe not directly an environmental problem as such but clearly also leading to an end to a civilisation that relies on natural resources no longer available. Often noted as a separate concern, but driving all of the above and climate change, is the exponential population growth of the last century, because more people means more environmental problems caused by people.

This chapter will look at these issues together, as they generally tend to be clustered together under the umbrella of environmental concerns, and they do all have a common origin in that they are consequences of human action and, more specifically in some accounts, the consequence of human scientific and techno-logical civilisation (the problem of climate change is of course also one of these issues, but because it has grown so big and important as a concern, I have decided to treat it in a separate chapter).

Of course, the same can be said of nuclear war, the topic of my previous chapter, even though unlike nuclear waste this is not generally held to be an environmental concern. Why is that? While I don't think this is in itself a particularly interesting question, it might help lead us into one more fundamental problem we might have to tackle in order to make sense of our language about the destruction of nature: What do we really mean by nature, and what marks nature out as different for our apocalyptic concerns?

The precise nature of what we mean by "nature" is very much contested. In the same way that introductions to the sociology of religion or science tend to start by stating the difficulty in defining religion and science, so some textbooks on the sociology of environmentalism start with complicating for us the term "nature" (e.g. Sutton, 2007). First, there are the two conventional meanings of nature, as shown by my entirely unoriginal pun in the chapter title. Nature can refer to either the essential essence of something, or to anything that is not associated with humans or human interference. Although most work on environmental sociology will naturally (sorry) focus on the latter definition, these two are related, and the former meaning can give us an indication of why pinning down the latter is often so difficult. As Sutton (2007, p. 2) points out, these two meanings are in some ways the extreme ends of the development of the concept, with nature first in the 14th century becoming the "forces, or indeed the force, that directed the world and ultimately explained why things happen when they do" (p. 2), still captured in the scientific sense of natural forces and natural laws. By the 17th century, nature came to be the "whole material world of things" (p. 3). This then leads finally to the more contemporary concept of nature from popular environmental thought that hinges on its separation from humanity, that is that there is an essence to the way the world exists that is independent of human interference. Nature is thus exemplified by the untouched rainforest rather than the urban concrete landscape. But as with many concepts that are defined in opposition to something else, there are problems, particularly if the nature (sorry again) of *that* is also not entirely clear. These can most easily be seen at the margins, for example the farmland landscape that conceptually sits between the rainforest and the urban concrete, so while it is in some way an oppositional, binary concept, the notion that some things can be more (or less) natural than others has also entered contemporary views on nature.

Thus, second, the concept of nature is complicated by being in one sense a binary, static idea and in another one, a matter of degree. How much human inter-vention is permissible in a landscape for us to still be able to call it "natural"? It is also further complicated when we try to dwell on the essential naturalness of humanity itself, we are after all products of natural evolution, and in that sense our products, our tools, cars, microplastics and haircuts can be argued to be as natural as the strawberries in my garden (which are in fact not quite so natural after all, as they are the result of centuries of selective breeding – not that the slugs seem to care). Where does nature end and humanity start, and is this even a useful question to ask?

This point on the uncertainty of what we mean by "nature" can easily be over-laboured of course, and all introductions to environmental sociology eventually move on to meatier topics. Just like with science and religion, we can take a vague and less defined common-sense understanding of the topic as our starting point and go on from there as long as definitional complications are acknowledged at the start.

However, in the story of how the environmental crisis relates to millennial imagination, the definitional relation between nature and humanity becomes a focal point. The first lecture of Latour's Clifford lectures (2017) concerns just this "instability of the (notion of) nature", again in an introductory context to his whole argument, but this time also as a theme in its own right. The imaginary of nature, as an Edenic paradise undisturbed by human intervention, underlies much of the nostalgia behind at least some strands of esoteric and environmental millennialism, which as a narrative strand connects the human scientific/technological interventions that threaten the environment conceptualised as the Edenic nature with the acquisition and utilisation of forbidden knowledge that exemplified the Genesis version of the Prometheus myth. As the historian of New Age and esotericism Nicholas Campion (2016) has noted, "A sense of the present crisis is invariably associated with the memory of a lost Eden or nostalgia for a vanished golden age, when the world was perfect" (p. 18).

Environmentalism and spirituality: parallel histories

Just as the link between science at large and belief is a larger area of controversy that nevertheless rarely finds itself in introductory textbooks on science or religion, so the link between environmental science and religion has similarly not been a much-discussed topic within the sociology of the environment. Sociological introduction to environmentalism (such as Yearley, 1992a; Pepper, 1996; Jamison, 2001; Irwin, 2001; Sutton, 2007), maybe especially those written from the perspective of Science and Technology Studies, rarely devote extended attention to religion. However, as I will try to outline here, the connections between the philosophies and ideologies that enabled developing environmental concepts, of nature, of our relation to it, its intrinsic and extrinsic values, and of course (in the context of this book) ideas about the causes and consequences of the *end* of nature, are all deeply embedded into a society's overall background myths, moral values and social knowledge. For better or worse, these are all bound up with religious and spiritual worldviews that have consequently coloured and developed our perceptions of nature, the environment and what it at stake in trying to protect it, even when the spiritual or religious belief has been lost itself (similar to how Weber's 1992 [1930] influential argument on the influence of Protestantism on capitalist society does not mean that modern-day capitalists need themselves be religious).

There does exist, however, a parallel area of scholarship on nature and religion that rarely finds itself discussed much in the science studies end of environmentalism studies, with some exceptions such as the recent work of Bruno Latour, which I will draw on here (published for example through the *Journal for the Study of Religion, Nature and Culture*).

This neglect is probably influenced through an influential view on the environment/religion relation, related to the "conflict" thesis in the wider science/religion debates, that religion has an overall harmful influence on nature. Very influential in establishing this view was Lynn White's (1967) article in *Science* echoing earlier arguments made in the tradition of Feuerbach (Taylor, 2016), which argued that Western religious traditions have (along with scientific innovations) largely been responsible for the widespread environmental damage that was becoming apparent by the 1960s. This argument was based on Christian anthropocentric beliefs on man's domination over the animals, as well as the "disenchantment" of a monotheistic tradition that moved spirits and minor deities out of our surrounding nature. Thus, White argues that "we shall continue to have a worsening ecologic crisis until we reject the Christian axiom that nature has no reason for existence save to serve man" (White, 1967, p. 1207). However, as a devout Christian himself, White's argument was one for reform rather than rejection of Christian beliefs. Despite its large influence, Whitney (2015) argues that there is now a consensus among historians and theological scholars that White's arguments have "oversimplified complex topics" (p. 403).

But spirituality/religiosity and environmentalism are more fundamentally intertwined (see e.g. Taylor, 2010). This is partly through more traditionally Protestant notions of "the book of nature" that developed as part of the philosophical and theological innovations that enabled the scientific revolution (Shapin, 1996), and which in a way found itself reasserted in the 19th-century US with the early national parks movement that conceptualised nature as God's temple (Stoll, 2015). But parallel to that, there were also developments in the new esoteric religious movements such as Helena Blavatsky's theosophical society and its many and varied intellectual descendants (Hanegraaff, 2013), which like 17th-century Protestantism also respected science (at least their interpretation of it) and had its own scientific adherents (von Stuckrad, 2014).

A reasonable start for a history of environmentalism and the modern concept of nature often would be the Industrial Revolution (e.g. Sutton, 2007). We only really started appreciating the natural world when it started moving away from us, when more and more people started to move to the cities and lost nature as an ever-present part of everyday life. Before the Industrial Revolution, nature was more likely to be seen as dangerous environment where wild animals and diseases lived and thus a place that needed to be "dominated" in the Christian tradition, as argued by White (1967).

One part of the reaction towards industrialisation can be seen by movements that nostalgically looked back to simpler times, as Campion (2016, p. 18) noted a move back to the lost world of Eden, or an imagined golden age; this reaction showed itself both in more secular, artistic movements such as the Romantics but also as noted in the spiritual worlds such as American Protestant denominations (Stoll, 2015) and the developing esoteric theosophical tradition (Campion, 2016; Hanegraaff, 2013). But philosophically, these movements also had roots in more (by then) established enlightenment traditions and even earlier that somewhat preceded the Industrial Revolution. Rousseau's notion of the early, natural society of men, which itself was a development from earlier renaissance philosophies such as the lost innocence of the "perfect state of nature" that Montaigne lamented (Campion, 2016, p. 28). These concepts would later reassert themselves in the Romantics' notions of the "noble savage". And of course, the myth of the garden of Eden and general nostalgia for the passing of golden ages predates the Industrial Revolution as well.

However, the late 18th and particularly the 19th centuries, when the process of industrialisation speeded up, can be seen to be very important in the development of any modern sense of environmentalism. My emphasis above on the nostalgic and Romantic element towards the loss of nature links the birth of environmental consciousness to the spiritual, millennial nostalgia that looks towards the future, that the world we lost can be regained again.

Industrialisation is of course also a logical starting point for the history of environmentalism because it was the process of industrialisation itself that caused most environmental problems. Other than drawing people into towns and cities and thus stimulating the nostalgia for nature lost, industrial pollution, the burning of fossil fuels and the immense population increases of industrialising societies are all directly the originators of most of our current environmental concerns. This is not to say that humanity had no negative impact on the natural world before (although, as outlined above, the "natural world" in the sense I mean it here would be an anachronous term for pre-industrialised society); as Diamond (2011) has pointed out, for example, even apocalyptically disastrous environmental degradation has been depressingly commonplace pre-industrialisation. But maybe our instinctive nostalgic reaction towards pre-industrial but still perfectly "unnatural" rural farm landscapes as idyllic betrays the origins of our environmental concerns as part of industrial society's nostalgia for a simpler, rural way of life.

In any case, the industrialised and industrialising nations by the late 19th and the early 20th centuries saw the birth of various societies and organisations that dedicated themselves in some way, not just to nature as abstract construct but to nature and/or nature interests as a way of reconnecting to that nostalgic of a lost, simpler life.

In Germany, nature conservation movements sprang up alongside parallel movements such as hiking and rambling associations (such as the *Wandervögel*) that romanticised nature and a back-to-nature lifestyle often combined with a clear nationalistic outlook. These movements included a wider German back-to-nature lifestyle fashion that was in the process of developing and finding its outlet in increasing popularity of vegetarianism, abstinence, naturism and more. This is often included in the umbrella term of the *Lebensreform* (life-reform) movement, which developed in parallel and in connection with the esoteric spiritualist thinking of theosophy and later (specifically in Germany) anthroposophy advocating similar ideals (see below).

Broadly, however, the emerging concerns particularly on the environment that accompanied this trend were not a mass movement and were often confined to bourgeois interests. Such conservation movements developed out of the "reactionary critique of modernization" which "had its roots in much earlier intellectual critiques [of German Romanticism] of the Enlightenment's emphasis on rationality, individualism, materialism, and the levelling of social distinctions and cultural differences" (Markham, 2008, p. 49). This socially conservative reaction was based on nature conservation rather than focussing on issues that affected quality of life, such as pollution. Developed and championed by intellectuals such as the musician Ernst Rudorff, this early conservationist environmentalism was intimately tied up with nationalist images of a "German national character as rooted in its intimate connections to an organically evolved, harmonious and interconnected folk culture and the German landscape" (Markham, 2008, p. 50), and these sentiments were to be a precursor of the environmentalism of later German nationalist movements (see also Rohkrämer, 2002, on how the German environmental past still influences its contemporary Green movements).

While environmental notions had also been articulated from socialist left-wing thinkers, a broader movement never took off, with socialist activism being devoted mostly to more urgent priorities, while environmental conservation was seen as a part of bourgeois culture. Thus left-wing environmental intellectuals workers' broader participation in environmental organisations could "divert attention from party and union activities" (p. 53).

These movements, when they organised themselves into specific clubs or organisations, were at first also largely local affairs, probably reflecting the strongly decentralised nature of the German state that had only recently emerged out of unification. Of the national organisations, Markham highlights in particular the *Bund Heimatschutz* for who Rudorff was an activist; the *Naturfreunde*, a workers' nature appreciation society; the *Bund für Vogelshutz* (League of Bird protection); and the *Bund Naturschutz Bayern* (Bavarian League for Nature Protection) – these were all focussed on conservation/protection and nature appreciation, organisations that focus on pollution were notably absent.

Parallel but also distinctly different developments occurred in other industrialising nations at the same time. The early environmental movements in the US, similarly focussed on issues highlighting nature (and particularly bird) conservation, with for example the Audubon Society (Greenwood, 2007). The Sierra

Club, co-founded by John Muir, a Scottish-American naturalist, was instrumental in the establishment of national parks which combined hiking and conservation elements.

Just as in Germany, though, the beginnings of this environmental movement were enabled by romanticist critiques of industrialisation and its effects; similarly, it was a more bourgeois rather than a mass movement. The romanticist reaction and critique towards industrialisation took different intellectual forms, one based on the different, more diverse, but also more puritan religious heritage of that country, which also then manifested itself in a different intellectual flavour behind the early environmentalist movement. One of its inspirations was the strong transcendentalist tradition that inspired what Bron Taylor (2010) labelled the "dark green religion" of contemporary spiritual environmentalism, which emerged through the work of Ralph Waldo Emerson and Henry David Thoreau, among others (and was itself inspired by German romanticism), who argued that spiritual truths could be found directly in nature, which in turn inspired John Muir. All three of these writers have written what Taylor called "almost sacred" texts for later deep ecologist thinkers (Taylor, 2010, p. 57).

As Stoll (2015) outlines, early environmental and national park movements also derived from the intellectual heritage of the various Protestant churches that became mainstream of American religious thought, such as Congregationalism and later Methodism, Presbyterianism and others. Muir himself had a Scottish Presbyterian upbringing. Thus, while a similar romanticist heritage influenced the beginning of US environmentalism, its Protestant heritage in the US had spiritually similar effects in terms of the reaction towards industrialisation and the accompanying rapid social change.

In Germany, the conservative and conservationist element of environmentalism as it had developed by the early 20th century made it into a natural part of the conservative and nationalist mix of ideas that made up the developing philosophy of National Socialism. The environmental/conservationist themes of a nation unspoiled by modernity and evoking images of a past golden age naturally appealed to the not always entirely coherent Nazi ideology and slotted in easily into their millenarian restorative vision of a thousand-year Reich of peace and prosperity (for the Germans at least). As has often been remarked, Hitler himself was both an advocate for animal protection and a vegetarian (Markham, 2008, p. 72), the latter of which shows that Nazi ideology has also been influenced by the *Lebensreform* movement that looks nostalgically back to a simpler life (Biehl and Staudenmaier, 1995).

The birth of modern environmentalism as a progressive critique of technological progress and its effects is often dated to the 1962 publication of Rachel Carson's *Silent Spring*. Already a well-known science communicator, Carson described the devastating effects of industrial and agricultural pollution have on wildlife. The title evokes the silence of a spring that would otherwise be alive with the sound of birds. The impact of Carson's book can be credited to her deft combination of technical and technically sound scientific critique and the urgent register through which she expressed herself:

> We stand now where two roads diverge. But . . . they are not equally fair. The road we have long been traveling is deceptively easy, a smooth superhighway on which we progress with great speed, but at its end lies disaster. The other fork of the road – the one "less travelled by" – offers our last, our only chance to reach a destination that assures the preservation of the earth.
>
> (Carson, 1962, p. 240)

Carson's rhetoric – and the effect of that rhetoric – has been described as prophetic by Walsh (2013), who devotes a whole chapter in her book on *Scientists as Prophets* to Rachel Carson. Carson's personal tragedy may also have contributed to the prophetic character – she was dying of cancer at the time she wrote her book and died shortly after its publication, and most of the impact and discussion about *Silent Spring* happened posthumously. With cancer being one of the illnesses associated with the unregulated use of industrial and agricultural chemicals Carson was writing about (though ultimate the cause of her illness may of course have nothing to do with it), Carson also ended up having become something of a martyr to environmental pollution. While this is not to say that Carson herself was motivated by spiritual concerns, her deft and poetic use of language as well as the circumstances of the publication of the book arguably helped find a wider audience it might otherwise not have.

Another pivotal development of the 1960s, coming to a head a few years after the publication of Carson's book, was the rise of the counterculture. An emerging youth subculture saw an increased resistance to the establishment, even if the 1960s have by now been somewhat mythologised themselves (Campion, 2016). The '60s youth movements had different and more complicated roots and expressed itself in different ways. In the US this was partly through resistance and opposition to the Vietnam War; in France it expressed itself through a student rebellion that was more politically left wing than the stereotypical hippie movement in the US; in Germany it manifested itself as the coming of age of the post-war generation that began asking their parents uncomfortable questions about how they could have let National Socialism happen, or even what *they* were doing during the war. But even in one country the counterculture was composed of different groups and identities, united with an overall feeling of generational resistance and in many ways shared cultural expressions, but nevertheless with often different views and philosophies on the detail, different points of origin and different trajectories after the initial period.

On the spiritual side, much of the counterculture was inspired by an esoteric heritage linked to theosophy and earlier religious spiritualist beliefs and subcultures. These beliefs were, like the counterculture itself, not homogenous or in any way a unified system, however they represent a line of influence that inspired the spiritual beliefs of many participants.

One of the strong underlying facets of this heritage was a specific type of millennialism, as exemplified through the name "New Age" – while this was not a new term, it came to be clearly identified as a wider connotation of the spiritual beliefs of the counterculture that sprang to prominence in the 1960s. New Age refers to the ideas that humanity stands at the cusp of a new age, an age that is more enlightened and in often less defined ways, "better"; there are clear analogies and conceptual relations here with the Christian millennium and wider utopian imaginaries of the better future. Another term, again older than 1960, was that of the Aquarian Age. This is an idea inherited from astrology that postulates the earth to be moving into a new era every roughly 2,000 years due to the precessional rotation of the earth's axis; the Age of Aquarius is 1 of 12 parts of a "great year", each 2,000 years long, evoking alongside the millennium of the New Age the tradition of dividing history into several distinct ages in Zoroastrianism, the prophecy of Daniel and more recently dispensationalism.

Ideas of the New or Aquarian Age had a wider pop-cultural influence on how the 1960s were remembered afterwards and became a wider part of the spiritual inheritance of a variety of new religious and quasi-religious movements of the 1970s and after. Marilyn Ferguson's (1980) influential book *The Aquarian Conspiracy*, for example, envisioned society to be on the cusp of becoming better, more attuned to nature and spiritually evolved. It needs to be noted that these were not always consciously religious; Ferguson herself claims her work to be based on sound philosophy and science, and the Aquarian of the title was not given any clear astrological significance either. However, as I outlined in Chapter 2, on the margins these boundaries are not always easily drawn. Ferguson's views of the evolution of human society and human spirit to something qualitatively new and better, also resonates with and was inspired by the theology of Teilhard de Chardin (1959), who argued that humanity will evolve to an "Omega point", a final unification of everything.

This New Age millennialism, while clearly part of the Western, Christian millennial tradition, also differs clearly from it, owing to its esoteric heritage, and the specifically 1960s and '70s borrowings from South and East Asian religious traditions – it is a progressive millennialism that welcomes the transformation but also in most forms does not envisage a catastrophic, apocalyptic transition (Campion differentiates between millennialism and apocalypticism for that reason, though as noted in Chapter 1, these two terms are being used in different ways by different authors).

The apocalyptic as a catastrophic future can be found in the adoption of environmental concerns within the new religious movements, which now combines the conservation and back-to-nature romanticism of traditional environmental movements that had already been influenced by the esotericism of the 19th and early 20th centuries with the acute, scientifically formulated concerns raised by Carson and the new environmental movement she inspired.

The scientific (non-countercultural) environmental movement that developed after Carson achieved a wider recognition with the 1972 report *Limits to Growth* (Meadows et al., 1972), commissioned by the Club of Rome. As the name indicates, this report was in particular concerned with the projection of economic growth into the future, arguing that growth cannot continue indefinitely. The concern over unlimited growth echoed broader ideas that were beginning to achieve wide currency in the 1970s about human population growth (which of course also links to economic growth). The human population had by then already reached 3 billion and appeared to be increasing exponentially, meaning that at some point the capacity for the planet to support human life will be exhausted. This projection was not particularly far off, since human population has continued to grow rapidly since then; it is now greater than 7 billion (United Nations, 2019). The concern is by no means new and is usually traced to Thomas Malthus' *Essay on the Principle of Population* (1992 [1798]), which argued that since human population increases exponentially and our capacity for producing food only linearly, at some point there will be a problem with potentially catastrophic results. This line of thinking experienced a renaissance in the 1970s, not just with the more economically focussed Club of Rome report but also as a separate, widely received best-selling work by the biologist Paul Ehrlich's *The Population* Bomb (1968): Pepper (1996) labelled the message of "ecocentrism" of the 1970s as "starkly neo-Malthusian" (p. 65). Concerns over the population "explosion" have since died down considerably, owing partly to the fact that early projections about the carrying capacity of the planet appear to have been rather pessimistic and it is still unclear what its actual limits are, but also because these concerns can easily transition into more unsavoury xenophobic concerns since the argument can easily be converted into the concern that there are too many "others" rather than of us.

The *Limits* report received a lot of criticism with respect to the date and methodology it used (as outlined in Pepper, 1996, p. 68), with the authors themselves noting in an updated study in 1992 that their conclusions constitute "merely a conditional warning" (p. 68) and that the resource depletion they warned about is no longer directly imminent. Its political and social impact in shaping the agenda of the new political environmentalism of the 1970 was big, however, and was followed by the very influential report commissioned by Gro Harlem Brundtland, *Our Common Future* (usually referred to as the Brundtland report; World Commission on Environment and Development, 1987).

Beyond the uptake of environmental concerns by the United Nations, the era saw two other developments that helped move environmentalism into the political arena, first as a marginal set of movements and later as a mainstream agenda: Environmental protest groups and green parties. The environmental protest groups were spearheaded by groups like Greenpeace (founded in 1971) and re-emerging established environmental conservation groups (Jamison, 2001, p. 92). Other groups such as Friends of the Earth (1969) and WWF (1961) also emerged or gained in prominence, alongside many smaller and local ones, and more radical groups such as Earth First! (1980): All had slightly different modes of operation, philosophies and agendas, but worked towards similar goals of drawing attention to, and thus politicising, the urgency of the environmental message.

The path of politicisation of the environmental message was taken more directly by a number of green parties that emerged in the late '70s and early '80s, with varying levels of success in the European countries. Their trajectories mostly depended on the various political systems and internal strategical choices between the wings that wanted to emphasise drawing attention to environmentalism and therefore staying clearly on message, and the wings that emphasised electoral success but also compromise as the best way of promoting the cause (Jamison, 2001).

The different fortunes of the green parties in the UK and Germany illustrate how local political frameworks shaped the fate of green politics. The electoral system in the UK is notoriously difficult for small and special interest parties to make electoral headway, and such parties on the whole therefore tend to concentrate on message purity, with the party being a platform for putting visibility on their agenda rather than affecting direct legislative reform. A similar effect also influenced green party politics in the US. By contrast, the federal and proportionate representative system in Germany made early electoral success possible, first on the state level and later on the federal level. By the 1990s, the Green Party had positioned itself as the main alternative progressive left-wing party in the country and became part of the governing coalition in Gerhard Schröder's 1998–2005 government. This favoured the compromising voices in green politics and also forced the party to take political positions on non-environmental topics, which further entrenched green politics as a broadly left-wing voice, giving them larger legislative influence but at the price of positioning environmentalism into already established political narratives; this would later on make it more difficult to make environmental arguments part of right-leaning and conservative thinking. And so, of course, the success of the green campaign movements and the green parties also depended on how you define it.

Thus, by the mid-1980s, the environmental movement, never of course a clearly monolithic movement to begin with, had developed into many various factions with different philosophies, modes of operation, ideas of what is at stake and solutions (and whether the solutions would or should be based on scientific-technical, social or even spiritual innovations). Added to these broadly political and campaigning sides of environmentalism, the spiritual heritage of the 1960s new age millennialism also coloured a lot of the emerging environmentalist movements, with esoteric and spiritual discourses emerging about nature and our role in the world in relation to nature. This strand of green politics is illustrated for example by the career of David Icke, who was a Green Party spokesman in the late 1980s but then moved on to become a cult and conspiracy theory leader (Robertson, 2016).

All these – the scientific/technical, the campaigning/political and the spiritual/religious elements of environmentalism – are not clearly delineated into different warring groups but rather represent different poles in a broader dispersed movement often at odds with itself. Certainly, many on the scientific side of environmentalism would claim to be atheists, while others are members of mainstream churches that don't have much natural sympathy for new age philosophy.

But the spiritual side of environmentalism, on the other hand, does not claim to abandon the science – on the contrary, environmental science has been built into a wider spiritual worldview (or rather, sets of often conflicting worldviews), reflecting the fact that new age spirituality as a whole never had a real bone to pick with science and instead incorporated what it found useful. Ferguson's *Aquarian Conspiracy* is filled with references to science, and philosophers of science such as Kuhn are deemed to be an essential part of the argument. This includes science that may not be entirely orthodox or might be pulled out of context, of course – new age arguments on expanding consciousness may refer to the quantum physics of David Bohm and the chemistry of Ilya Prigogine, for example (Hanegraaff, 2007).

One of the most interesting scientific ideas that has greatly influenced the spiritual side of environmentalism – and even some of the more political and scientific environmentalist thinking – was the Gaia hypothesis by James Lovelock (1979). Lovelock's main idea is of the planet/nature as a self-regulating organism. This is not to say that the earth is an organism in the traditional sense, even less so that it is a conscious one. Instead, Lovelock makes a comparison to a thermostat, a system that can tolerate small disruptions to the equilibrium by adjusting itself. In a similar way, an individual animal or plant may tolerate and accommodate small variations in their local environment. The Gaia hypothesis has struggled to gain acceptance from the scientific establishment, which at best gave it a grudging tolerance as a poetic metaphor, although the force of this pushback has died down somewhat over the years and its main ideas are being taken more seriously (Turney, 2003; Latour, 2017). It has in any case been influential as part of a wider philosophical consolidation between the spiritual and the scientific sides of the environmental movement.

Because of environmentalism's politically ambivalent history with national socialism, ecology is however also still a feature of contemporary far-right politics. This phenomenon – dubbed "ecofascism" by Biehl and Staudenmaier (1995) – links back to ecology's roots in nostalgia for lost and simpler times that National Socialism mined to build into their wider ethnocentric ideology. In a process that Biehl labels as "hijacking", contemporary neo-fascist ideologies have therefore also taken environmentalist ideas as part of their belief system:

> In ways that sometimes approximate beliefs of progressive-minded ecologists, these reactionary and outright fascist ecologists emphasize the supremacy of the "Earth" over people; evoke "feelings" and intuition at the expense of reason; and uphold a crude sociobiologistic and even Malthusian biologism. Tenets of "New Age" eco-ideology that seem benign to most people in England and the United States – specifically, its mystical and antirational strains – are being intertwined with ecofascism in Germany today.
>
> (Biehl and Staudenmaier, 1995, p. 3)

As they note, ecofascism has not merely incorporated environmental ideas into their thinking out of political opportunism when these views became increasingly mainstream in the 1990s, but also because some of its traditional links include both esoteric "New Age" millennial spiritualism and the Malthusian worry over population explosions. These were easily incorporated into their fear of a racialised "other" intent on outbreeding and thus replacing Western, white, communities. This resulted in a racial and environmental apocalypse that has gained in popularity in the contemporary conspiracy theories associated with the rise of the "Alt-right" and popular authoritarianism of the 2010s (Forchtner, 2019), for example through the "White genocide" conspiracy that motivated the 2019 Christchurch shooting (Moses, 2019).

Connections with racist and far-right ideas have plagued more mainstream environmentalism as well, mainly through the Malthusian link that leads worries over overpopulation to morph into worries about immigration and the environmentally taboo-breaking behaviours of the usually racialised outgroup. It is also linked through Arne Naess' influential concept of "deep ecology" (Drengson, 1995), an environmental philosophy that argues for the intrinsic value of nature outside of its utility to human beings, which in some interpretations can easily end up relegating human needs as secondary. The human needs that can be disregarded in favour of nature then easily become those of the outgroup and again, the racialised other. As Blair Taylor notes, deep ecology appeals to the "mystical and esoteric orientation of the alt-right" because of "a shared antimodernist politics and preference for changing culture over statist solutions" (Taylor, 2019, pp. 277–278). This can be seen in some of the acrimonious splits in environmental activist groups such as Earth First! in the 1980s. Earth First! co-founder Dave Foreman, for example, argued against "aid to famine-stricken Ethiopia so that Nature could 'seek its own balance'" (cited in Stoll, 2001, p. 412). Comparisons and allusions to environmental language are further made in contemporary alt-right literature, where white Europeans are presented as the equivalent of indigenous wildlife and immigrants as invasive species, with white Europeans facing "habitat loss" (Taylor, 2019, p. 278).

Mother nature and her prophets

The conceptual connections that environmentalist thought shares with parallel religious and spiritual developments is not to argue that environmentalism is a religious or spiritual issue as such. Instead my point would be that religious and spiritual beliefs of various flavours are an integral part of our cultural background that shape, influence and generally interact with our social perceptions of other concepts, even if individually we might not hold to any of these beliefs. In this, environmentalism would not be alone, of course, and as I tried to show in the previous chapter, the religious awe that the nuclear bomb inspired in almost anyone who contemplated it was also not necessarily confined to adherents of any particular faith. Just as the bomb forces us to consider the role of evil in our new secular eschatology, so does environmentalism force us to confront the potential end of the world (Buell, 2004; McNeish, 2017), which then inevitably gets framed in the apocalyptic terms our cultural background knowledge is already familiar with. With this in mind, we can try to identify several anchor points for an apocalyptic interpretation of environmental crisis.

Nature as God

One of the popular appeals of James Lovelock's concept of Gaia (and one of the reasons some scientists instinctively disliked it) is that it can be interpreted as a deification of nature: Lovelock chose to name his concept of the surface of earth as a superorganism after the Greek goddess of nature on suggestion of his novelist friend William Golding (Lovelock, 1979, p. vii). This name choice was not a meant as a deification, but the rhetorical effect was nevertheless there.

In his discussion of Lovelock's Gaia, Latour (2017) has noted that the explanatory usefulness of the concept of Nature as used in much of scientific thinking can be transposed to and understood as a similar explanatory device to the concept of God in various religions. Latour's own scheme is more nuanced than simply equating Nature with God, however in a simple intuitive way we can readily see how this could work, and this simple intuitive way is often how scientists who have not given this more direct thought might approach this themselves. Appeal to God as the primary cause, following Aristotle, is the ultimate explanation beyond which we cannot go further. The similarity of Aristotle's prime mover with appeal to Nature, at least in popular scientific accounts, can sometimes be striking. Clearly, Nature does not "abhor a vacuum", or anything else, because it is not a living agent itself and thus not capable of feelings like abhorrence. In this sense Nature has become an anthropomorphisation which is used as a convenient metaphor for popular scientific explanations. Of course, in no real sense is current scientific thought really attributing agency to Nature and thus making Nature into God. However, this this also presupposes attributes to God Herself, which may not be universally relevant to religious understanding of God. While a non-agent God may be a bit too much of a stretch in the concept of God, a generally non-intervening God might fit the bill.

The philosopher Nancy Cartwright makes a similar point, in that she argues that "the concept of a law of Nature cannot be made sense of without God" (Cartwright, 2007, p. 1), although she draws the conclusion that the laws of nature might need to be abandoned as a concept in science. Seeing that at least in regular scientific discourse the concept of the laws of nature is still very much alive despite Cartwright's argument, a conclusion that God does still play a role in scientific explanation is presumably justified. In any case, the argument can be made that, while in contemporary scientific thought Nature is not equivalent to any traditional concept of God as such, they both still fulfil a similar explanatory function.

Next to Nature fulfilling (at least some of) the explanatory functions previously ascribed to God, it can also be argued that in some of the newly emergent New Age religions, Nature has come to also fulfil some of the emotional aspects of God. The idea of God is appealing emotionally when She is depicted as a benign loving father who looks out for us, cares about us; similarly Nature has come to be seen within the more spiritual circles of the environmental movement as benign, caring and reassuring, which may be one of the reasons Lovelock's Gaia idea had the success it had despite scientists' misgivings. My facetious subversion of God's gender above in this case also points towards Nature having taken on some of the forms of the "mother goddess" familiar in some non-Abrahamic religions. This may however represent an equivocation of the concept of Nature, as Nature in physical science explanations of natural laws is a different beast from the Nature of the spiritual environmental and New Age movements.

Ambiguity, though, is one of the key mechanisms through which ideas travel to new communities. According to Moscovici's theory of social representations (Moscovici, 2001, outlined in Chapter 4), new, difficult and/or abstract objects are socially being made sense of through anchoring them to our background social knowledge, norms and values. These representations are firmed up through the process of objectification, when the previously abstract and new concepts are being made concrete. The concept of Nature, then, despite the evolving nature of its meaning and its clearly abstract nature (sorry), has objectified as a concrete object in environmentalist writing, being further personified in concepts like Gaia, just like Nature was personified in older religious and spiritual contexts such as "Mother Nature", or the original Gaia goddess – or, indeed, God the Prime Mover Herself.

Scientists as prophets

Walsh (2013) has provided a rhetorical analysis of prophecy, and how the processes that prophecy uses to persuade to the truth of its knowledge claims can also be seen in the work of prominent scientists and science popularisers (see also Caduff, 2014, for an anthropological take on scientists as prophets in the context of pandemic preparedness). That rhetorical analysis is key to understanding how science seeks to persuade should be clear, because no scientific theory or result, no matter how important, will get anywhere if the relevant scientists do not manage to persuade their peers to take it seriously. The "truth", logical consistency or ability to explain known evidence of a scientific result or theory are all extra-rhetorical attributes we might think a science needs to possess, however the appeal to these characteristics is by itself a rhetorical manoeuvre, which can and will be used by prophecy as well. Alongside appeals to these measures, successful persuasion also often relies on the charismatic properties of the scientist or prophet, the emotional appeal the message has for the audience, and the circumstances through which the message gets relayed. Following Aristotle (2012), classical rhetoric scholarship divides these appeals into logos (appeal to the content of the message), pathos (appeal to the emotions), ethos (appeal to the character of the messenger) and kairos (appeal to the circumstances).

That these categories and more modern rhetorical concepts can be used to analyse scientists has been explored by a considerable literature on the rhetoric of science and popular science (e.g. Ceccarelli, 2001). The contribution that Walsh has made was to look specifically to the rhetorical practices of prophecy (e.g. the Oracle of Delphi) and trace them through the main scholars of the scientific revolution such as Bacon and into modern popular scientific discourses of Sagan,

Gould or Hawking. In this sense, then, a comparison can be made of contemporary, charismatic scientists fulfilling the (rhetorical) role in society that prophecy has been playing previously.

It could possibly be argued that conventionally at least, prophetic appeal has rested largely on charismatic authority (ethos), and prophetic charisma has emerged with Weber as one of the main strands of analysis on prophecy (see also O'Leary's [1994] further elaboration of prophetic charisma). By contrast, it could be argued that scientific appeal rests largely on appeal to the message (logos). Traditional research on the sociology of science following Merton (1973) or the grand narrative philosophical theories of scientific method would follow this lead since these analyses tended to concentrate on the message itself, relegating pathos, ethos and kairos to the explanations of why things go wrong in science. However, as research within Science and Technology Studies has shown fairly conclusively, scientific theories gain recognition not just on the grounds of their content. They also depend on the characters and credential of the scientists, the social practices they engage in to further their career, the time and place in which they appear, and even their emotional appeal to the public and other scientists.

Similarly, then, it would also be possible to look at prophecies through the rhetorical appeals more familiar to science. Prophetic appeal to logos might be unfamiliar as a way of looking at prophecy. However, clearly a prophecy's fortunes will also depend on the message and how the message itself stacks up. This would be in the form of logical consistency, confirmation or falsification by the evidence and so on. A prophecy that predicts circumstances which don't come to pass (something particularly pronounced in apocalyptic prophecies since the world is still very much there) needs to react to its failure, and the way it does so can be analysed in a parallel way to how we analyse scientific failure.

Environmental redemption, environmental dispensations and the environmental millennium

The religious requirements to live a simpler, usually more rural, life that disparages (much of) modern technological innovation have a long history. The new technologies of industrialising society (Blake's "Satanic Mills") could easily be associated with evil and sin, as they disconnect us from the now crystallising concept of God's nature. This is echoed in the Blumenberg's analysis of the myth of Prometheus which warned of the dangers of knowledge and technology. This association found its way in religious, spiritual and artistic expressions, but also in the secular world, with for example Mary Shelley's decidedly secular interpretation on the Prometheus theme (Turney, 1998). Thus at the time of the birth of modern environmental movements, a culture-wide association of new industrial life and technology and evil coloured segments of society of different faiths and none.

The injunctions to (often rural) simplicity that were offered as remedies – by traditional and faith groups, the Romantics and the transcendentalist environmental campaigners they inspired, and the developing secular environmental movements – can be seen to offer personal redemption to community and faith members. By removing oneself from the evil of the modern world, even if only in the small and symbolic ways, one can save oneself. Redemption not just as a Christian theological concept, but as a broader cultural memory even within a secularised and/or spiritualised society then also finds its way in how secularised environmentalism easily morphed into the lifestyle choices of the *Lebensreform* movement. We find a very personal feeling of redemption by eating muesli instead of processed, industrialised food (Levinovitz, 2020).

None of this is to say that the industrial processes that cause environmental damage are not to be resisted in some way, or maybe even objectively evil (or of course, that personal redemption is a bad thing). However, this cultural memory of redemption might help explain the injunctions and priorities of some contemporary strands of environmentalism, especially those that derive from the more spiritual ends of either New Age or Protestant thinking, which highlight our environmental problem as a matter of personal choices often at the expense of more meaningful collective action against environmental harm.

Personal redemption for the environmental sin of modern life might be joined by the more distributed redemption of a section of society if it is felt that it is responsible in some way for the environmental damage. Clearly in need of redemption are the industrialists and capitalists, as well as other such evil forces, though these have increasingly come to play the part of the evil that needs to be transcended or vanquished as part of the environmental millennialism. An interestingly ambivalent position within green thought is played by science/technology itself (Yearley, 1992b): Science/technology is clearly one of the main culprits of the story – technological inventions such as the steam and later internal combustion engines are responsible for industrial pollution, as are the scientific discoveries that drove and/or refined the technologies. As noted by the sociologists of reflexive modernity (Chapter 6), the big, late modern globalised risks are mostly technological in origin, even if it was other social groups and forces such as capitalism and globalisation that made them into a problem. And yet, without the efforts of scientists like Rachel Carson and the many others who joined and work on the environmental sciences, we wouldn't even know about the dire environmental situation we find ourselves in; furthermore scientists are actively working on technological solutions to many environmental problems. In that vein, science and technology are clearly not as beyond redemption as (at least for some of the radical groups), capitalism would be.

In this vein the calls, often articulated from within science, for new ethically responsible ways of doing science have started with the dawn of the risk society in the 1960s. The UK "radical science" movement, for example (Rose and Rose, 1979; see Bell, 2017), called for democratically accountable and socially progressive science in the wake of the nuclear fears of the (science-led) cold war. But more recently, sometimes rather vague calls for "interdisciplinarity" as a way of making science accountable and progressive or the citizen science movement that in some of its loftier rhetoric promises bottom-up, community-led science can be argued to be attempts at a redemption for science as a whole (Prainsack and Riesch, 2016; Riesch, forthcoming). This is of course not to say that a more democratic, socially responsible and progressive science is not desirable, so none of this is a criticism of either citizen science or interdisciplinarity, nor of the goals that these are often seen to promise. However, seeing these movements as attempts to redeem science explains why these more progressive forms of science are seen particularly as solutions within environmental science (particularly so for citizen science), even if the actual mechanism for how exactly a more progressive science should be more successful in solving environmental issues is more often taken as read rather than formulated precisely.

One of the more striking imageries that have been developed as a way of visualising and dramatising the environmental apocalyptic is the concept of the Anthropocene (Steffen et al., 2011; Lewis and Maslin, 2015). The idea behind the Anthropocene is that human interference in all aspects of the environment has been so pervasive and permanent that it will be permanently visible – there has been a qualitative shift in the environment that qualifies our time as the beginning of a new geological epoch. While this proposal is by no means scientifically uncontroversial, the rhetoric behind the Anthropocene has managed to capture the wider public imagination in a relatively short time (Haraway, 2015).

The *Journal for the Study of Religion, Nature and Culture* has devoted a special issue on the religious elements of the Anthropocene, with Sideris and Whalen-Bridge (2019) noting its grand narrative about the place of humanity within the cosmos (see also Danowski and Viveiros De Castro, 2017). In this vein, there are similarities between the Anthropocene concept and age-old apocalyptic concepts that divide world history into clearly separate ages – the Zoroastrian ages, the metallic ages of Daniel, the dispensations of the Pentecostals, the Age of Aquarius – with us, our generation, always being at the cusp of the transition of the last age into the apocalyptic final time. My argument would then be that maybe it is the (Western) cultural memory (which I argued also influenced comparisons with Promethean evil of technology, personal redemption, the secular environmental and nuclear prophecy) that has helped propel the not yet scientifically accepted image of the Anthropocene so neatly into wider public consciousness.

Finally worth mentioning here, the environmental apocalypse is also a flavour of millennialism – not in all instances a catastrophic millennialism (in Wessinger's terminology) with a pessimistic outlook towards humanity's role but also often a progressive one, where environmental collapse is being viewed as a chance for society to build a fresh and just new environmental Jerusalem (as outlined in Chapter 7; Garforth, 2018).

Chapter 10

Climate apocalypse and the nature of prophecy

Introduction: climate wars and paradoxes

On the day I signed the contract for this book in 2019, I was sitting in my office in a T-shirt and students were sunning themselves on the lawn outside. Disturbingly though, this was in February, and this day would go into the record books for being the hottest-ever February day in Britain. However, this was a rather short-lived record, as the following day turned out to be even hotter (BBC, 2019). This unseasonal heat wave of course made the news at the time, and climate change, for a very short while at least, became the topic of the day. But it is also slightly worrying how quickly it had been forgotten about again – unseasonal, record-breaking heatwaves have over the past decade started to appear with alarming frequency, and in 2019 Britain at least, the national conversation would soon return to the big issue of the day, Brexit. In any case, I thought this was a fittingly symbolic day on which to commit myself to writing a book that is at least partly about climate change.

A year later, with Brexit having temporarily faded from the national conversation, another series of record-breaking heatwaves in spring and summer was completely overshadowed by the coronavirus crisis. Climate change is a pressing problem with disastrous and far-reaching consequences, but concerted international action, as well as a wider public recognition of its urgency, is still disappointingly meagre, probably because the consequences are hidden in a less immediate future and rarely relevant to everyday life. Formulated with characteristic modesty by the sociologist Anthony Giddens (2009) as "Giddens' paradox", the problem is that

> since the dangers posed by global warming aren't tangible, immediate or visible in the course of day-to-day life however awesome they appear, many will sit on their hands and do nothing of a concrete nature about them. Yet waiting until they become visible and acute before being stirred to serious action will, by definition, be too late.
>
> (Giddens, 2009, p. 2)

As Smith and Howe (2015) wryly remark in the introduction to their analysis of climate change as a "social drama", speculation like Giddens' on why the climate crisis is not being taken as seriously as scientists tell us it should be has become something of a "social science cottage industry" (p. 2).

Interestingly enough, comparisons with how we *should* react to this world-wide crisis are often made through the example of global pandemics, including by Smith and Howe: "This failure [to change lifestyles or engage in collective behaviour in response to climate change] is remarkable when contrasted with other global threats, notably, in response to terrorism, nuclear proliferation, toxins, or virus outbreaks like SARS and swine flu" (2015, p. 68). This argument clearly hasn't aged very well, since the experience of COVID-19 (as of autumn 2020 at least) has been that a large section of society does not appear to take viral outbreaks particularly seriously either. Speaking of nuclear proliferation, more than half a century earlier Bertrand Russell also used the threat of pandemics and our expected more rational reaction to them as a rhetorical contrast to how we should deal with the threat of nuclear war (2010 [1959]).

So, in some way it seems to be a case of the grass being greener for other apocalypses: We imagine society behaves singularly irrational when confronted with the crisis we personally worry about most. But the experience of COVID-19 appears to indicate that we don't do so well when confronted with any other worldwide crisis either, so there must be something more to all this than the diagnosis that Giddens has given us regarding climate change; global warming is not as unique a problem in terms of public reaction as Giddens thought. This is all, of course, somewhat worrying and disappointing.

And yet, as I've tried to outline in the rest of this book, our societal imaginaries and worries about a catastrophic near future are pretty much constant and a large part of our cultural heritage. In closing this book with a look at climate change, I will try to use my earlier musings on other apocalypses to try and find an answer to this problem (which, if I had any of Anthony Giddens' self-confidence, I might label "Riesch's paradox"): If the threat of an imagined imminent catastrophic future is something that has constantly worried people throughout the previous centuries, how come are we so relatively unconcerned when the catastrophe really does appear to be imminent?

Nordhaus and Shellenberger (2009) talk about "apocalypse fatigue", and this might be an appropriate label for the fact that centuries of (failed) predictions about the imminent end of the world may have conditioned us to complacency when the world really is in danger, so in a way we as a society have been caught crying wolf too often. While I believe there is some insight in this argument, it does however leave the question of why this particular apocalypse (rather than any of the others) should be the one that breaks the villager's caution about the wolf.

Alternatively, historical apocalypses show that overall, the majority of the villagers have never really been overly worried about the wolf. Apocalyptic and millennial movements are often just that – movements that are part of a wider society – but more rarely do they encompasses the whole society; even at its height, Millerism encompassed just a fraction (albeit large) of American society. Apocalyptic movements are outsiders, often define themselves through their status as outsiders, and a large part of their rhetorical force derives from that outsider status.

Being outsiders, their apocalyptic visions are not taken seriously by many – a complement to our long history of apocalyptic predictions is that they're always viewed sceptically by the rest of us. However, the story is more complicated, and groups and movements overlap and interact in complex ways. "Believers" in the science of climate change tend as a whole not to believe in the apocalyptic scenarios of UFO cults or dispensationalist Protestant churches, and it is an often remarked feature of conservative, evangelical communities (such as the dispensationalist congregations) that they either don't believe in climate change or that it is an urgent problem (Carr et al., 2012) – the stereotypical "denier" community. However, the overlaps are there, and the implications and interrelations of who believes what are constantly shifting. Some of those worried (or excited) about a biblical apocalypse being within the near future also believe that climate change is one of the "signs" of the coming apocalypse that was foretold in scripture (Green, 2014), in a similar way to the expectations of thermonuclear war being the "fire and brimstone" of Revelation.

Theoretically the fact that CO_2 can act as a greenhouse gas that could change the planet's climate has been known for a long time, with the (possibly) independent discoveries of John Tyndall and Eunice Newton Foote in the 19th century providing the mechanism, and Svante Arrhenius demonstrating how the burning of fossil fuels could alter earth's climate (Weart, 2008; Bell, 2021). Climate change became more than a theoretical event by the late 1950s when Charles Keeling first started producing ongoing measurements from the Mauna Loa Observatory in Hawaii showing that CO_2 concentrations in the atmosphere are rapidly increasing. By the late 1970s the concept of global warming had been more firmly established within scientific circles, and it has also started to be taken up as one of the wider environmental concerns that became increasingly prominent after Rachel Carson. With continuously improving climate models and temperature measurements, the basic facts of climate change have been fairly undisputed scientifically since the 1980s. Through its association with environmentalism that itself had firmly attached itself as a broadly left-wing political movement by the 1980s, it also became part of the wider political division that enabled conservative politics to dismiss it (alongside other environmental concerns), despite the fact that several conservative politicians (notably Margaret Thatcher in the UK) did take the message from climate science very seriously.

The last 30 years, however, have seen the problem become increasingly prominent in public discussion. The UN's Intergovernmental Panel on Climate Change (IPCC) has since 1988 provided regular reports on climate change, its risks and possible responses. Al Gore, former US vice president and prominent environmental campaigner, produced a concerted effort in the early 2000s to bring climate change into wider public attention through his popular documentary and tie-in book *An Inconvenient Truth* (Gore, 2006). In the UK, the influential Stern review (Stern, 2007) made the economic case for action on climate change, arguing that it will be cheaper in the long run to address the issue. At the same time, wider international political efforts to coordinate a response to climate change have been painfully slow. Despite occasional positive agreements in Kyoto (1997) and Paris (2016), these have also been criticised at length for not going far enough and interspersed with more sobering developments such as the widely criticised Copenhagen Summit (2009) or President Trump's decision to withdraw from the Paris Agreement.

While international public (Poushter and Huang, 2019) and overwhelming scientific (see e.g. Mann, 2014) opinion has stabilised around recognising the urgency of the climate crisis, powerful counter-narratives have slowed progress and become part of the general worldview of a large section of the international public, which varies strongly from country to country, with these knowledge communities being particularly large in the US and to a lesser extent the UK, compared to other European and Western countries. This is largely the result of a deliberate campaign to promote doubt and scepticism about climate change conducted by vested interests such as oil companies. Oreskes and Conway's (2011) account clearly demonstrates of how public relations strategies to "manufacture doubt" that have been developed by tobacco companies have been adapted successfully to propagating doubts about the science of climate change. But just because there is a cynical motive behind much large-scale public anti–climate science communication, this should not mean that many views that regard climate change science with scepticism are not sincerely held. Opinions about climate change, its severity and whether it is a problem at all have become part of a wider set of issues that have become part of the ingroup norms, values and background knowledges of a wider political identity. We are currently seeing the politicisation of attitudes towards COVID-19 into a similar, wider culture war; climate change has been part of this culture war for a far longer time and in that sense has been even more directly baked in to group identity. In this situation, new evidence, new information or attempts at scientific education about climate change will easily be re-interpreted in a way that accords to the general group-defining norms and values (Chapter 5), with the new evidence only managing to change peripheral assumptions.

In a similar vein, relating the way cultural affiliation affects how we interpret risks and scientific information, the social psychologist Dan Kahan (departing from Douglas and Wildavsky's 1982 cultural theory of environmental risk; see Chapter 6) notes the phenomenon of "cultural cognition" (Kahan, Jenkins-Smith and Braman, 2011; Kahan, 2012) as a way of explaining different groups' differing interpretations of the climate evidence. The vicious attacks on climate scientists that Mann (2014) recounts, for example, or the notorious 2009 hack of the "Climategate" emails (Grundmann, 2012) all underline the enormous amount of emotive force climate change (and opposition to it) has created by having become a part of a wider social identity. This emotive work is also evident on the other side, with epithets such as "climate deniers" (evoking parallels to Holocaust denial) further embedding the animosity between the two sides. With this entrenchment of climate change as a social and political identification issue, a resolution to what Mann calls the "climate wars" appears a long way off.

However, away from the science or larger arguments about how we through our social and political identities understand and mentally process the scientific arguments, I have also tried to argue in the two previous chapters that a wider system of cultural representation is present in our meaning making of globalised risk stories that lean on our cultural background understanding of apocalyptic and similar arguments. In this sense we can make comparisons between the wider apocalyptic narrative tropes that I have outlined previously, and how we talk about climate change. This, however, is complicated by the fact that to a much greater extent than the two sets of issues from the previous chapter, climate change is disputed, politicised and an ongoing locus of contention. At the same time, the links with the religious feelings of the "sublime" that was and still is present in discussion about the awe- and dread-inspiring destructiveness of nuclear weapons does not seem to be as much part of public representations of climate change – nor does the spiritual, religious streak of environmentalism infuse climate change debate as much as it does other or more general environmental issues. While maybe this might provide for a tentative additional explanation for why climate change as an existential risk is more contentious than these other topics, it is worth noting that while attempts to link climate change to apocalyptic narrative tropes are continually being made, these have been somewhat superficial (Skrimshire, 2014), and climate change as a disaster also offers additional challenges that make this link more difficult to establish. Climate change is a gradual disaster that, unlike the suddenness of the apocalyptic scenario creeps up on us over generational time scales (at least in the early days); it is a complex and abstract issue that is more difficult to fit and translate into more directly threatening, traditional disaster scenarios of war, famine, fire and brimstone.

Apocalyptic discourses and narratives

Hulme (2009) influentially argued that one of the reasons why people disagree about climate change is due to our background beliefs and values, highlighting the importance of spiritual and religious value systems in framing our interpretation and response to this crisis. The apocalyptic as a discursive frame in public discussions on climate change has been widely noticed (e.g. Fava, 2013; Swyngedouw, 2013; Anshelm and Hultman, 2015), and the apocalyptic as a narrative frames how we bring our background knowledge and value system into how we think about it. This may have some advantages on framing the crisis as an urgent issue for the catastrophic near future but also brings with it some potential pitfalls.

Visions of destruction: floods, wars, brimstone and famines

One particular way through which the dangers of climate change can be made more concrete as well as linking it to traditional apocalyptic narratives is to highlight the particular dramatic consequences through which it may cause, if not the end as such, at least widespread destruction. In Chapter 4 I noted floods as one of several apocalyptic visions of destruction that appear periodically in various apocalyptic vision. Worldwide flood, through the widely anticipated (and already observed) sea-level rise that climate change will bring, has become one of the main talking points through which the dramatic effects of climate change are being discussed. The vision of New York or London being flooded, or most of the Netherlands, is an enduring image of much popular climate change communication (Weingart, Engels and Pansegrau, 2000; Höijer, 2010).

Sea level rises are however only one part of a whole set of issues that climate change will cause, and while catastrophic news for low lying countries like the Netherlands (and even more so for Bangladesh and the Pacific Island nations), there are other destructive consequences to climate change. One of them is the widespread famine that would be caused by making already marginal farmland in areas such as the African Sahel completely unproductive, and therefore occasioning potential resource wars, large scale migrations and the all the attendant problems from that (see the widespread analysis in Stern, 2007). This further raises the spectre of war as one of the traditional apocalyptic consequences of climate change.

These issues relate to other directly relevant apocalyptic tropes, though they are also a lot less concrete in that they rely on a chain of events that are less directly attributable to climate change as a direct (and only) cause. Famine, of course one of the traditional interpretations of one of the horsemen of the apocalypse, is also one of the recurring images of climate change (Manzo, 2010). The 2020 wildfires in Australia and the West Coast of the US are being publicly interpreted as consequences of climate change (e.g. Borunda, 2020) and thus provide a link between climate change and the fire-and-brimstone trope.

Signs and portents of the hour

The portents of the coming end are all around us and are part of most climate change communication strategies – the arctic ice is forming later than usual, a little bit later every year, the glaciers are melting, and of course every unusual heatwave and abnormal weather event signifies that the end is coming. Wearing T-shirts in London in February, apocalyptic wildfires, record-breaking hurricane seasons, melting glaciers – all these are dire portents. None of this of course is to say that these are not clearly things we should be worried about, nor that they are not caused by climate change. It is also not to say that the public discourse should not focus on these as the signs and warnings that climate change really is happening and that it really is something we should be worried about. In fact, our tendency to focus on these events makes a lot of sense in that it does concentrate minds and helps persuade that things really are getting serious.

However, the rhetorical impact and discursive functions that these signs and portents of coming doom have are the same for climate change as they are for any of the other apocalypses that have never happened or turned out not to be something to worry about. Tabor (2011, p. 256) notes signs and portents of the end as one of the general features of apocalyptic prophecy after Daniel: It is precisely because signs and portents are so persuasive towards belief in the coming end that they have featured so heavily in previous predictions and prophecies, and to this fact the veracity of climate change makes no difference. This, like the points below, does however allude to the strategies that climate prophecy must take if is to be persuasive, and conversely this might help convince us why previous prophecies were so powerfully persuasive that apocalyptic predictions have been a recurrent feature of (at least Western) civilisation. At a simple, folk-epistemological level, the prophecies make predictions, and while the main event stays elusive and therefore cannot be used to corroborate (or falsify) the belief, the smaller, less important predictions can be seen to come true, thus rendering confirmation to the overall belief system. As Robertson (2016) notes with his concept of "rolling prophecy", the phenomenal success rate of corroboration in conspiracy theory and apocalyptic prophecies is explained by the sheer number of predictions that are made, of which some will end up coming true. Here the difference with the scientific apocalyptic of climate change is that there is no "rolling prophecy", in the sense that science overpredicts and then cherry-picks those naturally occurring events that confirm the belief and downplays all the other non-fulfilled predictions. However, the difference is more likely one of degrees, as historical and sociological studies, and philosophical reflections like that of Lakatos have shown, overprediction and disregarded evidence to the contrary does occur in science frequently. While in science there are measures that guard against this, these are not foolproof, and moreover from inside the belief system it is difficult to be a competent and impartial judge on this. The upshot is that the signs and portents feel the same to those invested in the relevant belief and fulfil the same rhetorical function in entrenching our beliefs.

Dates and urgency

One of the most prevalent components of the apocalyptic imaginary is that the coming end is imminent, another general feature of apocalyptic prophecy that Tabor (2011) listed. Predictions of the end have traditionally been accompanied with precise dates for when the event will happen; all these precise predictions have come and gone without dampening the enthusiasm for them in the long run. Individual apocalyptic movements do however learn from the failure of precise dates, and as the example of the theological change from Millerism to Seventh Day Adventism has shown, such movements can change the date into an undetermined future after the initial refutation(s). Yet this also appears to signal a move away from the apocalyptic into mainstream religion, especially if a movement itself becomes part of the wider establishment that the initial fervour sought to overcome. In any case, in order to be a convincing apocalyptic narrative, the future cataclysm needs not just to be continuously corroborated through signs and portents but also to have a clear and decisive end point that is near enough in the future to be worrying.

This poses a problem for climate change communication, since the catastrophe will occur gradually and almost imperceptibly until long after we have crossed the point of no return – this is Giddens' paradox. The signs and portents help in focussing our minds, but in order to be truly worrying, a close-enough end date needs to be in sight. And many climate science communicators have duly obliged in providing some semblance of certainty and convictions in providing these dates, sometimes with the "point of no return" standing in for the single cataclysmic event that climate change will otherwise not produce. Addressing the European Parliament in April 2019, for example, the climate change campaigner Greta Thunberg predicted that

> around the year 2030, 10 years, 259 days and 10 hours away from now, we will be in a position where we will set off an irreversible chain reaction beyond human control that will most likely lead to the end of our civilization as we know it.
>
> (Thunberg, 2019, p. 46)

Thus Thunberg evokes the urgency of the issue through both referring to a point of no return and setting a clear and precise date for it – the abstract and nebulous event of climate change is made concrete and objectified, even though the precise-ness of the date is more a rhetorical embellishment than a real prediction. In fact, 10 years seems to be a fairly frequently used number for climate change warnings; see for example the press release to Hawkins and Sutton (2012; University of Reading, 2011) or a recent speech by Prince Charles (Daily Mail, 2020).

The clear dangers of these communication strategies are that, while they may be effective in the first instance, they lose potency if these dates pass without clear incident. It is also of course a distortion of the scientific evidence which is simply unable to give such precise dates (see Dunne, 2020 for a further discussion). In this, the gambit of using points of no return with precise dates will be possibly counterproductive when that point of no return has come and gone. Thus warnings made 20 years ago to the effect that we have only 10 years left will appear overblown and overdramatic to us now, even if the 10 years was merely used as a rhetorical device and the main point of the argument still stands. Or indeed, even if the message was accurate and we have already reached and passed the point of no return without noticing it: Since nothing much visibly happened, the prophecy is seen to have failed either way and – like Festinger's cults – merely moved the end a little bit further into the future or into an indefinite future, saving the belief system at the cost of its credibility from the outside.

Instead of invisible points of no return, the apocalyptic prediction can instead look towards particularly real catastrophic events instead. While most of climate change happens gradually and is therefore unsuitable for precise date-setting, some large-scale catastrophic events – such as floods, abnormal hurricane seasons or unseasonal heatwaves – can fit the bill, but since they do happen often enough without precipitating global catastrophes, they are not cataclysmic enough. A lot of attention has therefore focussed on the public discussion of "tipping points" (Lenton et al., 2019), or single events that accelerate the process by a large amount and can therefore appear as clear and demonstrable future events that push us visibly beyond the point of no return. These tipping points, such as the loss of the West Antarctic ice sheet, represent moments where the system suddenly changes and becomes unstable and lead to large, qualitatively different changes. The rhetorical impact of tipping points is that of producing clearly identifiable and thus demonstrably catastrophic events for an otherwise gradual process (Horn, 2014).

The causes of climate change: Prometheus and climate devils

Most of the points I noted in the previous chapter about general environmentalism are also of course applicable to climate change. In particular, the personification of nature as the equivalent of a benevolent god or goddess that in general has become thoroughly secularised, through scientific (or quasi-scientific) concepts such as Lovelock's Gaia. Alternatively, we conceive of Nature as the non-human-influenced state the world would be in if it weren't for us, a counterfactual situation where Nature serves as an explanatory in lieu of earlier concepts of God.

Similarly, climate change, as an event caused by human technological intervention, evokes the Prometheus myth that infuses environmentalism in general. Part of the salience of Prometheus can in fact be seen in the dismissals of climate change that are often made on the climate "denier" side: An argument that concedes that the climate is changing, and that the consequences of this are just as destructive, but that the causes of this are not technological but entirely natural. The point is then often made as part of an argument that we should not intervene in climate change (despite the likely highly destructive consequences that would presumably be just as bad as if climate change was caused by humans). The reason this argument sounds so persuasive, despite appearing to be very relaxed towards catastrophic consequences, appears to be related to the missing element of the Promethean narrative here – without the human intervention in nature, without fire having metaphorically been stolen from the gods, and without forbidden knowledge having led to the destruction of Eden, what will happen will happen, and our likely fate is entirely natural and therefore tolerable or even good in itself (see Levinovitz, 2020, on this equation of Nature with goodness).

With climate change having become part of a politicised group identity on both sides of the debate, blame and responsibility are being attributed to the party deemed responsible for the current mess. Scientists as a group have escaped blame – the ambiguity in green politics towards science and technology that Yearley (1992b) wrote about means that while scientific and technological progress in the abstract are part of the Promethean bargain, it is now also largely scientists who discovered the problem, try to warn us about it and even do something about it. The blame instead goes to members of the outgroup as we would expect.

Depending on which particular green or environmental group we look at, these outsiders could be capitalism (both in the more abstract, for example in the groups that connect climate change activism with anti-globalisation thinking), or more concretely in the targeting of specific individuals or institutions (politicians like first George W. Bush and then Donald Trump, industrialists, newspaper tycoons). And again, this is not an argument that these individuals and their institutions are blameless in worsening climate change through either direct action or cynical attempts at manipulating public opinion against action and towards doubt. However, these demonisations are also clearly related to the wider ingroup/outgroup dynamic through which group and social identity gets constructed and maintained, with the attendant danger or side effect that perceived nefariousness in motivations and taboo-breaking conduct get accentuated while behaviour and views that accord with the ingroup norms are downplayed. This may not appear to be particularly unfair when we consider the altogether negative impact that individuals like Donald Trump or Rupert Murdoch have had overall, however the clear danger is that these mechanisms can work – and frequently do – when the evidence of nefariousness is much less shaky or even non-existent. For anti-capitalist and anti-globalisation campaigners, this may then be the industrial class and elite politicians with vested interest in the climate status quo, however this can easily morph into the projection of climate evil to other outgroups we have already been suspicious about. Conspiracy theories about a New World Order that have been bubbling along in the background for decades become reanimated through the new climate taboo-breaking behaviour of the global elites that these theories are worried about. Antisemitism has come back through precisely these channels, as the outgroup malevolence that endangers the rest of us, while this trend is part of a wider group of politicised cultural issues, environmentalism and climate change are also part of this (as shown in the previous chapter on "ecofascism").

Thus while a clear and demonstrable evil is a useful narrative through which to conceptualise and communicate the climate danger we are currently finding ourselves in, the pitfalls are that the evil party becomes whatever outgroup we have already been worried about, and who may not deserve this demonisation. And as we have seen through the example of other apocalyptic movements, the apocalyptic other has throughout history been a target of prosecution and violence (Wessinger, 2000, and with a particular emphasis on 20th-century millennial movements, including environmentalism, Barkun, 1996). So a climate change campaign needs to be aware of this, that demonisation tactics are clearly useful in making the vague, gradual danger of climate change into an immediate apocalyptic threat to be taken seriously, but that they can also backfire and seek out the familiar outgroup devils of whichever particular social group we are trying to reach.

Rupert Murdoch or the oil company executives that Oreskes and Conway (2011) write about in my view are worthy targets due to their demonstrably adverse behaviour and the clear effect their actions have had in worsening the crisis. Uncomfortably, however, even if individuals, organisations or wider social groups are deserved targets of any specific blame-targeting behaviour, focussing on them rather than on the overall underlying issues can have the effect of diverting actions from where action is needed onto righteous but ultimately futile moral crusades.

One possible example is public attitudes towards nuclear energy, particularly in Germany (possibly due to Germany's cold war heritage of being the most likely battlefield for any third world war). Due to nuclear energy's association with previously demonised devils in the nuclear industry and its association with nuclear war, any debate about nuclear energy being a potential part of a lower-carbon energy future is immediately cut off (nuclear energy, for example, was a comparison used to argue against the otherwise unrelated technology of carbon capture and storage in focus groups conducted by Riesch et al., 2013). This is not to say that the benefits of the low-carbon nature of nuclear energy outweigh its disadvantages, however the point is that this debate is effectively taboo within green political circles. In Corry and Riesch (2012), we make a similar argument trying to explain why environmentalist groups are generally so lukewarm towards carbon capture and storage technology as (at least a temporary) solution to lowering carbon emissions in fossil fuel power plants: Some groups, with perfectly good justification, see fossil fuel energy companies as the general enemy towards progress in the fight against climate change, and therefore any solution that includes them is immediately suspect. On a more ideological level, how we should respond to climate change depends on what we think the root cause of the problem is, and if we're too embedded in an ideology that sees, say capitalism, as the main issue (because capitalism is our general outgroup devil), then the solution becomes having to get rid of capitalism first. However, as much as capitalism may be at fault for climate change, this is an altogether different proposition than finding pragmatic solutions to the immediate problem of carbon emissions. The extent to which this is problematic, then, is not dependent on the scientific arguments regarding climate change and its urgency, but an ideological one where the answer may differ between various environmentalist groups as much as within the wider concerned but unaffiliated public.

Salvation, redemption and the prophets of climate change

The apocalyptic is a worldwide disaster and ultimately the responsibility of a variety of outgrouped others: the Beasts, the people they deceive, and the various hordes from the north and east in Revelation; the Jews in some of the medieval apocalyptic movements that Cohn (1957) wrote about; the sabotaging remnants of ancien régime in the French Revolution; the Jews (again) in Nazi millennialism; or the capitalists in some of the current climate change discourses (and specifically Jewish capitalists in the ecofascist strand of climate discourse).

On an individual level, though, we can achieve redemption through our own actions: Just like the redemption offered through righteous and clean living in the general environmental discourse noted in the previous chapter, climate change offers individuals the chance of salvation through the actions that everyone can take themselves (see e.g. the green lifestyle manuals analysed in Rohloff, 2012, 2019). While climate change action needs to come through coordinated, international political actions, much of the moral climate change discourse is focussed on individual actions we can all perform – justified through the argument that if we all act on it the disaster might be mitigated, which of course I am not disputing, this individual action discourse nevertheless takes on a moralising tone that is reminiscent of and fulfils similar rhetorical and narrative work to the personal redemption narrative that accompanies traditional apocalyptic discourse. While the world is doomed, the saintly chosen are spared the coming time of trouble. This may well be a useful strategy to gradually change world public opinion to the point that coordinated political action will have to follow, so again I want to emphasise that I'm not against this discourse as such. However, it is of course also clear that I alone will not, by limiting my personal carbon footprint, effect climate change meaningfully. I will however, feel much better about myself, and that is (if I'm honest) one of the reasons why I aim to participate in this drive as well. The danger, then, is of course that individual action can become a displacement for the coordinated political action that is actually needed; most of us will individually feel we have done our bit and will therefore face the coming disaster with a clean-enough conscience. This redemptive element to climate change action then functions in a similar way to the faith, prayer and general leading of a righteous life that is a common response to traditional apocalyptic expectations – these individual actions will not save the world from the coming catastrophe but will save the individual soul.

There are several examples of this redemption narrative that accompanies climate change discourse – the injunctions against flying, not owning a car (or at least using it too often), buying locally produced foods and more recently also the probably biggest individual contribution towards our carbon footprint we can make in a lifetime, having children. All these actions and more are usefully summarised in the concept of the "carbon footprint", a variation/development of Wackernagel and Rees' (1996) "ecological footprint".

As with some of the earlier points I was making in this chapter, I don't want to argue that all these actions are not worthwhile, and while I personally will not save the world by limiting my flying, it is certainly the case that if we all do this, a significant source of carbon emissions might be saved. And living an individually cleaner life is at least not going to hurt anyone. The bigger problem with this personal redemptive reaction towards climate change is that it becomes a displacement preoccupation that distracts from the wider issue and the big structural changes to 21st-century world civilisation that need to happen fast in order to prevent the worst effects of climate change. As with the earlier comparisons of apocalyptic discourses, my point here is that these discourses have their equivalent in various apocalyptic narratives that climate change for better or worse has inherited as part of our general cultural background. But the contrast between the personal redemption stories we tell about our own lives and how we live it contrast favourably to the Promethean evil that characterises the outgroup's risk taboo-breaking behaviour and which will bring chaos to the world. In this sense, redemption and evil are the two sides of the ingroup/outgroup coin which characterises general group identities, and it is through this perspective that I am worried about the polarisation of efforts to mitigate climate science into an ingroup/outgroup dynamic which will ultimately stifle the needed coordinated efforts from all sides to act on climate change.

Rohloff (2012, 2019) talks about "carbon temperance" in her analysis of green living guides, in analogy to the abstinence movements on alcohol, drugs and so forth as a discourse that imagines carbon-rich lifestyles as an addiction. Similar to addiction, this introduces carbon-rich lifestyles as a complex interdependence of personal responsibility and forces outside of our control, but like addiction the ultimate responsibility lies with the individual. The personal responsibility element is heightened by the denigrating talk of "eco-criminals" and "eco-deviants" for those who do not take part in carbon temperance. Low-carbon lifestyles have thus become a matter of strictly policed morality that she analyses as an instance of Norbert Elias' (1994) civilising process.

In the previous chapter I outlined Walsh's (2013) argument about the rhetorical similarity between science popularisers and prophets. Rachel Carson, particularly through her book *Silent Spring*, was one of Walsh's case studies. Another was the prophetic ethos of the IPCC. Walsh was concerned mainly with the rhetoric of the IPCC's language and the prophetic registers it uses as an attempt to persuade its readers to the seriousness of its message. In general, though, it appears that the prophetic impact that Carson has had in launching the general environmental movement has been lacking in the IPCC; climate change was still looking for its Rachel Carson until recently, and needs "ethical and charismatic authority" and "heroes and heroines (like Carson and [Wangari] Maathai) to defend civil society against dangerous enemies" (Smith and Howe, 2015, pp. 189–190). Despite their hard work and obvious merits, the IPCC does not seem to have stepped into Carson's shoes in terms of the public imagination.

Prophetic language is only part of the prophet's persona that makes them appear so persuasive. O'Leary (1994) expands on Weber's concept of prophetic charisma by arguing that the persuasiveness of a prophet's charisma is not just derived from the size of the audience but also from the nature of the audience. If the prophet manages to persuade his followers that his audience includes not only them but also God and/or the universe itself, then the size of the audience becomes cosmic, and the charisma he or she derives from it (from the viewpoint of an audience member) becomes immense.

But there are aspects of the prophet's persona that go beyond the purely verbal rhetorical strategies and their audiences. Hinted at by Weber but discussed in more detail by Cohn (1957) is the humble origin of the prophet and their laicity. In the sense that the prophet is usually arguing against the contemporary worldly powers which includes any established churches as well, the prophet is usually an outsider, a lay preacher like William Miller or an outright heretic, ascetic or hermit who is living his life outside of the established power structures. This gives him the needed special authority to condemn the establishment. The prophet is therefore also often an outsider, a tragic character, a martyr who seeks and receives no personal gain for their speaking of truth (note of course that this is the image they need to project as prophets rather than always the reality).

I speculated in the previous chapter that Rachel Carson's persona as a science communicator has helped establish her popular prophetic credentials – although a trained scientist, she was not really part of a clear scientific, university-based establishment. But probably more important to her prophetic charisma was the tragic facet of her life, the fact that she was at the time dying of exactly the type of disease that she was warning about. This is as I said speculation, but these circumstances shine a very much different light on Carson as a person than if exactly the same book had been written by a healthy, wealthy and politically connected university scientist.

It is this aspect of Carson's persona that neither the mostly anonymous and establishment scientists of the IPCC nor the former vice president Al Gore could hope to emulate. If climate change wants a prophet, then it needs to be an outsider figure without political connections or scientific credentials, a tragic figure who is overcoming personal adversity in order to spread the message but seeks no other personal gain.

The extraordinary rise of the Swedish climate change campaigner Greta Thunberg, who at age 15 initiated the school strike movement in 2018, appears to fulfil several of the associations noted above to give her a prophetic persona. Her age and her lack of political connections or influence provide the prophetic impression of an innocent child speaking truth to power; her Asperger's diagnosis also enhances the outsider status and provides a narrative of the prophet overcoming adversity in order to propagate the message. Thunberg openly discusses her Asperger's as well as her age ("I'm too young to do this"; Thunberg, 2019, p. 25) and thus weaves both these aspects of her persona into her overall rhetoric. The persuasiveness of the outsider as a messenger of truth maps onto some of the traditional scientific norms that Merton (1973) famously analysed. Merton proposes *universalism* as a scientific norm, the idea that a message needs to be judged on its own merit regardless of who proposes it – by putting emphasis on their outsider status, prophets and lay-preachers direct our attention to the message; they hint at the fact that their message should be evaluated aside from the professional ethos they otherwise lack. Being outsiders, they have nothing to gain; they are only interested in the truth. This in turn links to Merton's scientific norm of *disinterestedness*, that scientists and scientific institutions should act for the wider benefit of science rather than their own personal gain.

In this context, then, we can note that some of the discrediting strategies Thunberg's detractors use to eliminate the prophetic character – from accusations of her using her Asperger's to arguments that she does in fact have clear political connections through her parents (O'Neill, 2019; Fawkes, 2019). These appear to be direct attempts at deflating the prophetic persona, if only because neither of these attack strategies are engaging with the actual arguments. Linking Thunberg to millennialism and prophecy, as O'Neill does, is also intended to discredit her message as it suggests that superstitious millennialism is all that the message portrays. By making this link myself here, I don't of course intend to follow O'Neill in the conclusion that therefore the message itself can be disregarded; this does, however, highlight the pitfalls of using prophetic charisma in communicating the climate change message.

In any case, the message itself and its delivery demonstrate remarkable oratorial skills, and in addition to working the apocalyptic register of climate change clearly ("either we choose to go on as a civilisation or we don't"; Thunberg, 2019, p. 21), allusions to the wider, universal audience that O'Leary notes also appear through, for example, the invocation of future generations that are looking back at us wondering how we could have let this happen (p. 15). Invoking the judgement of history serves as an appeal to O'Leary's universal audience: "I ask you to stand on the right side of history" (p. 18).

Boucher (2019) argues that in general, depicting Thunberg as a prophet is risking "distorting the message" and that this has "opened the floodgates to all sorts of messianic theories". Whether or not Boucher's argument is justified, however, the wider narrative of climate change as an apocalyptic appears to be forcing Thunberg into this role, and it is certainly too early to tell whether this has indeed distorted the message or helped it reach a wider audience.

Chapter 11

The end is near!

All things must come to an end, and so too does this book. The world faces a series of "existential risks" that need to be taken seriously – or at least much more seriously than it appears they are. This for me is mostly so for the case of climate change, though as outlined, other researchers (such as existential risk researchers) also argue this for other risks such as artificial intelligence. Our general social understanding of these risks that shape how we react to them and how we can ultimately overcome them rests on the narrative strategies we employ to make sense of them, understand them and assign wider meaning to them.

As new or relatively new concepts – like most technological and scientific advancements generally are – we do so through linking, or anchoring what we have learned of these new concepts to the wider social and cultural frames and narratives that are familiar to us; abstract projections of future developments, filtered through generally unfamiliar (for most of us) scientific language and theorising, become objectified, made concrete through this anchoring mechanism in a way that lets us understand it all in familiar terms.

As a way of imparting meaning to these cold and unfeeling scientific projections, we embed these future projections into a narrative that recounts frightful events into familiar stories, of heroes and villains, them and us, purity taboos and taboo-breakers, sin and redemption and wider, larger cosmic struggles for control of the future. These narratives bring us, individually, back into the centre of the action, as Kermode (1967) noted; our intuitive understanding that the end is near narrates us into the pivotal, decisive point in time, standing at the apex of history. Framing current scientific fears about globalised risks as *existential*, even if it is scientifically justified, imparts a more primeval, apocalyptic narrative onto world events that we intuitively understand better than the cold, meaningless, nihilistic simple end of it all. And moreover, it flatters us, because we as individuals and through our wider social and generational group identities are special in the revealed scientific knowledge that we now understand but the taboo-breaking outgroups don't, that we stand on the side of good in the coming cosmic struggle, whether we prevail or not.

My overall argument, then, is that the narrative structures that we make use of in order to make sense of contemporary globalised late modern or existential risks derive, or very often derive, from our wider cultural background myths of the apocalyptic. Within this established narrative structure, there are many details and tropes that we can, with a bit of imagination maybe, easily map onto what is currently happening with regards to nuclear war, or climate change, or catastrophic biodiversity loss, or any of the other contemporary existential risks that we are facing. These include parallels we can draw on the nature of the destruction such as war, famine, floods or even the stars falling from heaven. They include a conceptualisation of time that compartmentalises it into discrete ages; signs and portents of the coming end; humble prophets that warn and messiahs that save us; judgement and redemption, cosmic fights between good and evil; and the quiet hopes for the millennial utopia afterwards.

The apocalyptic is a narrative structure that shapes how we make sense of current crises, though it is maybe also important to note that this is not the only narrative through which we can interpret current crisis – Anshelm and Hultman (2015) for example include the apocalyptic as a prominent, but not the only, interpretative frame in their analysis of Swedish climate change discourse. However, I still believe that the sheer forcefulness and frequency through which current crises are compared to the apocalyptic is striking and probably so familiar that we often fail to notice it. Swyngedouw's (2010) paper on the politicisation of climate change, for example, makes a pun on the movie *Apocalypse Now* in the title; a different pun on *Apocalypse Now* appears in a recent book by the former UK conservative party politician Oliver Letwin (2020), providing a semi-fictional account of how the country might deal with a collapse of the national grid.

But another way to show the enormous power of relating our existing apocalyptic myths and narratives to current events is that they also end up directly shaping how we approach each current crisis in various ways. Evangelical doomsday interpretation of political current events in the Middle East direct and influence these events as well, and they in turn influence the trajectories of the denominations making the links. And we can make similar observations about our apocalyptic interpretations of globalised existential risks like climate change or nuclear war: The direct apocalyptic framing of the dangers of nuclear war influenced a larger peace movement that had clearly apocalyptic imagery now to warn of the dangers of a next world war; it influenced the development of the mutually assured destruction doctrine that directed cold war politics; and it provided parts of the signs and portents that millennial faiths were seeing, shaping and directing these faiths into new paths.

With the apocalyptic frame come dangers and opportunities for those of us who are worried about climate change, environmental disasters or nuclear war. The opportunities lie in the fact that the apocalyptic frame can be utilised as a way of communicating the issue and drawing attention to its urgency and destructive potential (Spoel et al., 2008). But even if the apocalyptic is not directly pursued as a communication tool, the persuasiveness of the apocalyptic narrative itself as noted by scholars such as Barkun or O'Leary can almost automatically do the work. We are naturally receptive to apocalyptic messages for the variety of reasons they and others outline. This would include for example the conceptual security they bring to a world that has been plunged into chaos and societal disorientation through disasters, wars or economic crisis (Barkun, 1974). But there are also the features of the apocalyptic that relate it to the folk-epistemology of, for example, the rolling prophecies (Robertson, 2016) that in conjunction with the heightened attention to the signs and portents indicate the corroboration of the wider belief system. This by itself is a point that is not necessarily related to the apocalyptic nature of the rolling prophecy mechanism, but the apocalyptic enhances the salience of it. The persuasiveness of the prophetic message also plays a part here. Again, if looked at purely through the rhetorical lens of how the message is portrayed, we don't need to be apocalyptic in our warnings in order to be persuasive, however through the wider narrative link of the apocalyptic, the whole range of "universal audience" to the rhetorical force of the problem of evil (O'Leary, 1994) becomes available alongside the prophetic character of the humble outsider.

The apocalyptic frame in a way relieves science communication from having to consciously use a persuasive trick that the wider scientific ethos would find uncomfortable, because the intuitive narrative constructions we are automatically performing as part of the audience will do the job for them. We don't need to point out that Thunberg's persona fits the prophetic model; we can even heartily deny that this should be relevant, however if that's how she is being perceived by the wider audience, then there is nothing the science communicator can do to stop it. And if it helps, why should they?

But the problem is, of course, it might not necessarily help. Boucher (2019) warns that talking about Thunberg as a climate prophet may cloud or obscure the message. But there is more direct and more serious danger from the general apocalyptic frame as well. Nordhaus and Shellenberger's (2009) concept of "apocalypse fatigue" points to the dangers of the apocalyptic frame: "Rather than galvanizing public demand for difficult and far-reaching action, apocalyptic visions of global warming disaster have led many Americans to question the science". This fatigue may be due to the apocalyptic-ness of the visions but also to some of the other problems Nordhaus and Shellenberger discuss (such as political polarisation or the vague and gradual nature of global warming). But the association of climate change or other contemporary crises with the apocalyptic, which is in turn often associated with marginal doomsday cults (in the pejorative sense), then lends the crisis a feeling of unreality, that the fears are overblown just as those of the long tradition of doomsday predictions have so far always been.

So the possible mechanism through which something like an apocalypse fatigue could set in make intuitive sense: The end of the world has been proclaimed often and forcefully and yet it has failed to occur every time. Moreover, the patronising incredulity we, as scientifically minded secularists worried about climate change, might feel towards the endless procession of evangelical cults warning about the imminent end can just as well be applied by *them* towards *us*.

Just as Festinger noted about apocalyptic end-times predictions, when *we* predict a dire near future, this doesn't mean that when it fails to come to pass exactly as we thought, we will abandon the message. We will instead believe that we were wrong only on details. But from a non-invested point of view, this looks rather different. The catastrophic predictions of the Club of Rome report did not come to pass, and even though the overall worry we think was justified and it was just the detail that was wrong, the prophetic failure has been used extensively as a repudiation of the whole environmental movement. Hamblin (2013) quotes Alex Kozinski, a Reagan appointed federal judge: "With the benefit of hindsight, we know that the Limits to Growth was a bunch of hooey; virtually nothing the Club of Rome predicted with such alarm has come to pass" (p. 249). This was an argument that was used to ignore Al Gore's (1992) book (see also Buell, 2004).

Didn't scientists warn about the imminent end of the world after Hiroshima? There never was a third world war, and though nuclear weapons still proliferate in different ways and still pose a clear threat, the frequently made prediction of a direct nuclear confrontation between the US and the Soviet Union has been abandoned. Ehrlich and others warned us about the imminent population crisis at a time when there were "only" 3 billion people on earth; now with more than 7 billion, we appear to have proven that there was enough space for all of us after all. Again, the threat has not evaporated; it has merely shifted (and of course, humanity can't grow indefinitely). But as with the nuclear war with the Soviets, the actual dire prediction of an imminent catastrophe, maybe even accompanied with a clear date, never materialised. And so on – for every failed scientific apocalypse there are good reasons as to why they didn't occur as predicted and why the threat still remains there, but these explanations are then generally of the same nature as the evasions that doomsday cults take in order to explain away *their* failed apocalypses while still maintaining that the general message stays the same; it's only peripheral details that the disconfirming evidence has changed.

Or this at least it has to look like from the sociology of knowledge (both of religion and of science, as outlined in Chapter 2) point of view, because we need to take a neutral attitude towards the content of the knowledge claims and consider how the causes of these knowledge claims are of generally the same sort of type. In this case, because we stay neutral on the content, there appears to be an identity-based cultural cognition that evaluates the other side's apocalyptic warnings as superstitious for the reason that their failed predictions get explained away so easily. My argument would instead be, following Lakatos, that changing the peripheral assumptions to our prophecy/prediction is a perfectly rational thing to do, and we'd need to accord this rationality also to the doomsday cultists as they're working with the same intuitive epistemology.

As I tried to argue in Chapter 5, at least one hugely influential social-epistemological mechanism that has been worked out in philosophy of science to account for a rational reconstruction of the advance of scientific knowledge, while accommodating both the actual behaviour historians of science like Kuhn have noted towards disconfirming evidence (as well as the logical loopholes that philosophers like Quine have pointed to as a limitation of seeing science as a coldly logical mechanism, again with reference to disconfirming evidence), starts to look uncannily like the mechanism that Festinger identified in his UFO cult as a way for rescuing an apocalyptic belief system in the face of failed prophecy. There is a difference in degree – some epistemic group's beliefs are progressive and some degenerating, in Lakatos' terminology, but it's not a difference in kind.

But why should this be surprising? In the end, both scientists and UFO cultists are human beings endowed with similar faculties for reasoning, and the folk-epistemology that drives their religious belief is similar to that which drives scientists (who by and large do not have a formal philosophy of science education either and whose epistemology is therefore also driven by innate intuition, like for everyone else). This is only uncomfortable if we need to have a reason for why scientists' personal epistemology is superior to everyone else's, and as far as I know, both philosophy and sociology of science have abandoned this idea in favour of institutional justifications for why science is more reliable than other forms of knowledge making. And while institutional norms – peer review, organised scepticism, fierce competitiveness and so on – may drive scientists towards making far more reliable knowledge claims, the epistemic justifications and the logic behind them stay pretty much the same.

And it's not just the all too easily caricatured UFO cults and evangelists we may find easy to dismiss here, because the argument of failed prophecy comes from sources within the academy itself – sociologists like Furedi (2006) and philosophers like Bruckner (2013) also argue that the doomsday predictors have had such a bad track record that this indicates current fears around issues like climate change are similarly overblown. We cannot easily argue that these are silly people with deficient epistemological radars, as the (entirely wrong) intuition would be with UFO cultists or dispensational doomsday prophets.

But where, then, are we left in terms of which apocalyptic scenarios, if any, we should worry about?

At the end, we all in our own ways seem to believe that the world is in great peril, whether this is through climate change or through other religious or secular apocalypses. Luckily, as I tried to show in the previous chapters, there is enough interpretative flexibility for people to incorporate climate change (or the other contemporary existential risks) into whatever wider apocalyptic narrative other people may be invested in, and through that mechanism our various collective representations of the apocalyptic itself can become the boundary object that Science and Technology Studies scholars have argued eases communication between epistemic groups.

Whether this will at the end help us move past disagreements on how and whether to act on the existential risks we are facing is a rather different point. Next to despair and destruction, the apocalyptic narrative is also one of hope and renewal, redemption and conquest of evil; this unfortunately means that action against the apocalyptic may not be what some people actually want – nuclear confrontation can bring about the events that lead to the New Jerusalem, likewise climate change can finally topple the evil capitalist and help bring about a Callenbach-style socialist ecotopia. Or, interpreted from another angle, the apocalyptic frame suggests that this is just going to be another one of a long line of previous prophetic failures.

The danger of the apocalyptic narrative is, then, that it drags us into the realm of the familiar and the expected, it forecasts an inevitable future, one we could either dread or look forward to, but one we cannot avoid – or on the other hand, it lulls us into believing there is nothing to fear, because it has always turned out right at the end, that there is no apocalypse coming, and there never has been. Thus, because the apocalyptic narrative can make climate change appear as nothing to worry about, inevitable or even desirable, I believe a note of hope and optimism is the wrong one to take because it breeds complacency and familiarises us to the optimistic side of global catastrophe. Therefore, I'm taking inspiration from Billig (2005) writing about the hidden dangers of humour: While some authors writing about global existential risks like nuclear war or climate change like to end with an upbeat optimistic note, I don't.

The End

Bibliography

Adorno, T. W. & Horkheimer, M., 1997 [1947]. *Dialektik der aufklärung*. Frankfurt am Main: Suhrkamp.

Aldridge, A., 2013. *Religion in the contemporary world*. Cambridge: Polity Press.

Anderson, A., Allan, S., Petersen, A. & Wilkinson, C., 2005. The framing of nanotechnologies in the British newspaper press. *Science Communication*, 27(2), pp. 200–220.

Anshelm, J. & Hultman, M., 2015. *Discourses of global climate change: Apocalyptic framing and political antagonisms*. London: Routledge.

Arendt, H., 1963. *Eichmann in Jerusalem: A report on the banality of evil*. New York: Viking Press.

Aristotle, 2012. *The art of rhetoric*. London: Harper Press.

Asimov, I., 1979. *A choice of catastrophes*. London: Hutchinson & Co.

Attridge, H. W., 2002. The messiah and the millennium: The roots of two Jewish-Christian symbols. In: A. Amanat & M. Bernhardsson, eds. *Imagining the end: Visions of apocalypse from the ancient Middle East to modern America*. London: I. B. Tauris, pp. 90–105.

Augustine, 1972. *City of God*. London: Penguin.

Austin, W. H., 1970. Isaac Newton on science and religion. *Journal of the History of Ideas*, 31(4), pp. 521–542.

Avilés, M.A.R., 2000. The law-based utopia. *Critical Review of International Social and Political Philosophy*, 3(2–3), pp. 225–248.

Bainbridge, W. S., 1997. *The sociology of religious movements*. London: Routledge.

Balmer, B., 2002. Killing without the distressing preliminaries': Scientists' defence of the British biological warfare programme. *Minerva*, 40(1), pp. 57–75.

Barbour, I., 2000. *When science meets religion*. San Francisco: Harper.

Barkun, M., 1974. *Disaster and the millennium*. New Haven: Yale University Press.

Barkun, M., ed., 1996. *Millennialism and violence*. London: Frank Cass.

Barkun, M., 2013. *A culture of conspiracy: Apocalyptic visions in contemporary America*. Berkeley: University of California Press.

Barkun, M., 2016. Conspiracy theories as stigmatized knowledge. *Diogenes*. DOI: 10.1177/0392192116669288

Bashir, S., 2002. Deciphering the cosmos from creation to apocalypse: The Hurufiyya movement and medieval Islamic esotericism. In: A. Amanat & M. T. Bernhardsson, eds. *Imagining the end: Visions of apocalypse from the ancient Middle East to modern America*. London: I. B. Tauris, pp. 168–186.

Bauer, M. & Gaskell, G., 1999. Towards a paradigm for research on social representations. *Journal for the Theory of Social Behaviour*, 29(2), pp. 163–186.

Bauer, M. & Gaskell, G., 2008. Social representations theory: A progressive research programme for social psychology. *Journal for the Theory of Social Behaviour*, 38(4), pp. 335–353.

Baxter, S., 2008. *Flood*. London: Victor Gollancz.

Baxter, S., 2009. *Ark*. London: Victor Gollancz.

BBC, 2019. UK basks in warmest February day on record. *BBC News*, 25 February.

Beck, U., 1992. *Risk society: Towards a new modernity*. London: Sage.

Beck, U., 2007. *Weltrisikogesellschaft*. Frankfurt am Main: Suhrkamp.

Beck, U., Giddens, A. & Lash, S., 1994. *Reflexive modernization: Politics, tradition and aesthetics in the modern social order*. Stanford: Stanford University Press.

Beckett, L., 2020. "All the psychoses if US history": How America is victim-blaming the coronavirus dead. *The Guardian*, 21 May.

Bell, A., 2017. The scientific revolution that wasn't: The British society for social responsibility in science. *Radical History Review*, 127, pp. 149–172.

Bell, A., 2021. *Our biggest experiment: An epic history of the climate crisis*. London: Bloomsbury.

Berg, C., 2002. *Theologie im technologischen Zeitalter: Das Werk Ian Barbours als Beitrag zur Verhaltsbestimmung von Theologie zu Naturewissenschaft und Technik*. Stuttgart: Kohlhammer.

Berger, P. L., 1967. *The sacred canopy: Elements of a sociological theory of religion*. Garden City, NY: Doubleday.

Berry, D., 2020. Voting in the Kingdom: Prophecy voters, the new apostolic reformation, and Christian support for Trump. *Nova Religio: The Journal of Alternative and Emergent Religions*, 23(4), pp. 69–93.

Bibby, R. W. & Brinkerhoff, M. B., 1974. Sources of religious involvement: Issues for future empirical investigation. *Review of Religious Research*, 15(2), pp. 71–79.

Biehl, J. & Staudenmaier, P., 1995. *Ecofascism: Lessons from the German experience*. Edinburgh: AK Press.

Billig, M., 2005. *Laughter and ridicule: Towards a social critique of humour*. London: Sage.

Bishop, K., 2009. Dead man still walking. *Journal of Popular Film and Television*, 37(1), pp. 16–25.

Bloch, E., 1986. *The principle of hope*. Oxford: Blackwell.

Bloom, P. B.-N., Arikan, G. & Courtemanche, M., 2015. Religious social identity, religious belief, and anti-immigration sentiment. *The American Political Science Review*, 109, pp. 203–221.

Bloor, D., 1973. Wittgenstein and Mannheim on the sociology of mathematics. *Studies in History and Philosophy of Science Part A*, 4(2), pp. 173–191.

Bloor, D., 1991. *Knowledge and social imagery*. Chicago: Chicago University Press.

Blumenberg, H., 1979. *Arbeit am mythos*. Frankfurt am Main: Suhrkamp.

Borunda, A., 2020. The science connecting wildfires to climate change. *National Geographic*, 17 September.

Bostrom, N., 2002. Existential risks: Analyzing human extinction scenarios and related hazards. *Journal of Evolution and Technology*, 9.

Bostrom, N., 2013. Existential risk prevention as global priority. *Global Policy*, 4(1), pp. 15–31.

Bostrom, N., 2014. *Superintelligence: Paths, dangers, strategies*. Oxford: Oxford University Press.

Boucher, E., 2019. The dangers of depicting Greta Thunberg as a prophet. *The Conversation*, 12 December.

Boyer, P., 1992. *When time shall be no more: Prophecy belief in modern American culture*. Cambridge, MA: Belknap Press of Harvard University Press.

Boyer, P., 1994. *By the bomb's early light: American thought and culture at the dawn of the atomic age*. Chapel Hill: The University of North Carolina Press.

Bruce, S., ed., 1999. *Three early modern utopias*. Oxford: Oxford University Press.

Bruckner, P., 2013. *The fanaticism of the apocalypse: Save the earth, punish human beings*. Cambridge: Polity Press.

Buell, F., 2004. *From apocalypse to way of life*. London: Routledge.

Bulletin of the Atomic Scientists, 2020. *Current time*. [Online] Available at: https://thebulletin.org/doomsday-clock/current-time/

Burgess, A., 2019. Environmental risk narratives in historical perspective: From early warnings to "risk society" blame. *Journal of Risk Research*, 22(9), pp. 1128–1142.

Burkett, J., 2012. The campaign for nuclear disarmament and changing attitudes towards the earth in the nuclear age. *The British Journal for the History of Science*, 45(4), pp. 625–639.

Burris, C. T. & Jackson, L. M., 2000. Social identity and the true believer: Responses to threatened self-stereotypes among the intrinsically religious. *British Journal of Social Psychology*, 39(2), pp. 257–278.

Byrne, P., 1991. Religion and the religions. In: S. Sutherland & P. Clarke, eds. *The world's religions: A study of religion, traditional and new religion*. London: Routledge, pp. 3–28.

Caduff, C., 2014. Pandemic prophecy, or how to have faith in reason. *Current Anthropology*, 55(3), pp. 296–315.

Callenbach, E., 1975. *Ecotopia*. New York: Bantam.

Campion, N., 2016. *The new age in the modern West: Counterculture, utopia and prophecy from the eighteenth century to the present day*. London: Bloomsbury Academic.

Camus, A., 1972 [1948]. *The plague*. New York: Random House.

Carr, W., Patterson, M., Yung, L. & Spencer, D., 2012. The faithful skeptics: Evangelical religious beliefs and perceptions of climate change. *Journal for the Study of Religion, Nature & Culture*, 6(3), pp. 276–299.

Carrington, D., 2019a. Plummeting insect numbers "threaten collapse of nature". *The Guardian*, 10 February.

Carrington, D., 2019b. "Insect apocalypse" poses risk to all life on earth, conservationists warn. *The Guardian*, 13 November.

Carson, R., 1962. *Silent spring*. London: Penguin.

Carter, B., 1983. The anthropic principle and its implications for biological evolution. *Philosophical Transactions of the Royal Society of London: Series A, Mathematical and Physical Sciences*, 310(1512), pp. 347–363.

Cartwright, N., 2007. *No god, no laws*. [Online] Available at: https://sas-space.sas.ac.uk/963/1/N_Cartwright_God.pdf (accessed 21.01.21).

Cassam, Q., 2019. *Conspiracy theories*. Cambridge: Polity Press.

Ceccarelli, L., 2001. *Shaping science with rhetoric: The cases of Dobzhansky, Schrödinger, and Wilson*. Chicago: University of Chicago Press.

Chang, H., 2012. *Is water H_2O?: Evidence, realism and pluralism*. Dordrecht: Springer.

Chilton, P., 1982. Nukespeak: Nuclear language, culture and propaganda. In: C. Aubrey, ed. *Nukespeak: The media and the bomb*. London: Comedia Publishing, pp. 94–112.

Cimino, R. & Smith, C., 2011. The new atheism and the formation of the imagined secularist community. *Journal of Media and Religion*, 10(1), pp. 24–38.

Cohen, S., 1972. *Folk devils and moral panics: The creation of the mods and rockers*. London: Routledge.

Cohn, N., 1957. *The pursuit of the millennium: Revolutionary millenarians and mystical anarchists of the middle ages*. London: Random House.

Cohn, S., 2007. After the black death: Labour legislation and attitudes towards labour in late-medieval Western Europe. *The Economic History Review*, 60(3), pp. 457–485.

Collins, J. J., 2002. Eschatological dynamics and utopian ideals in early Judaism. In: *Imagining the end: Visions of apocalypse from the ancient Middle East to modern America*. London: I. B. Tauris, pp. 69–89.

Corry, O. & Riesch, H., 2012. Beyond "for or against". In: N. Markusson, S. Shackley & B. Evar, eds. *The social dynamics of carbon capture and storage: Understanding CCS representations, governance and innovation*. London: Routledge, pp. 91–108.

CSER, 2020. *About us*. [Online] Available at: www.cser.ac.uk/about-us/

Daily Mail, 2020. We've only got 10 years to save the planet, warns Prince Charles on the 50th anniversary of landmark speech he gave on plastic pollution and other threats to the environment. *The Daily Mail*, 18 February.

Daniels, T., ed., 1999. *A doomsday reader: Prophets, predictors and hucksters of salvation*. New York: New York University Press.

Danowski, D. & Viveiros De Castro, E., 2017. *The ends of the world*. Cambridge: Polity Press.

Davis, J. C., 1983. *Utopia and the ideal society: A study of English utopian writing 1516–1700*. Cambridge: Cambridge University Press.

Davis, J. C., 1993. Formal utopia/informal millennium: The struggle between form and substance as a context for seventeenth-century utopianism. In: K. Kumar & S. Bann, eds. *Utopias and the millennium*. London: Reaktion Books, pp. 17–32.

De Chardin, P. T., 1959. *The phenomenon of man*. London: Fontana.

De Cruz, H., 2019. Religion and science. In: E. N. Zalta, ed. *The Stanford encyclopedia of philosophy*. Summer 2019 Edition. [Online] Available at: https://plato.stanford.edu/archives/sum2019/entries/religion-science (accessed 21.01.2021).

DeGroot, G., 2004. *The bomb: A life*. London: Jonathan Cape.

Derrida, J., 1984. No apocalypse, not now. *Diacritics*, 14(2), pp. 20–31.

Diamond, J., 1997. *Guns, germs and steel: A short history of everybody for the last 13,000 years*. New York: W. W. Norton.

Diamond, J., 2011. *Collapse: How societies choose to fail or succeed*. London: Penguin.

Diken, B., 2009. *Nihilism*. London: Routledge.

Dittmer, J. & Sturm, T. eds., 2016. *Mapping the end times: American evangelical geopolitics and apocalyptic visions*. London: Routledge.

Dobson, M., 2016. Building peace, fearing the apocalypse?: Nuclear danger in soviet cold war culture. In: M. Grant & B. Ziemann, eds. *Understanding the imaginary war*. Manchester: Manchester University Press.

Dobson, M., 2018. Protestants, peace and the apocalypse: The USSR's religious cold war, 1947–62. *Journal of Contemporary History*, 53(2), pp. 361–390.

Douglas, M., 1966. *Purity and danger: An analysis of concepts of pollution and taboo*. London: Routledge.

Douglas, M., 1992. *Risk and blame: Essays in cultural theory*. London: Routledge.

Douglas, M. & Wildavsky, A., 1983. *Risk and culture: An essay on the selection of technological and environmental dangers*. Berkeley: University of California Press.

Doward, J., 2020. "Quite frankly terrifying": How the QAnon conspiracy theory is taking root in the UK. *The Guardian*, 20 September.

Drengson, A., 1995. The deep ecology movement. *The Trumpeter*, 12(3), pp. 2–6.

Dundes, A., ed., 1988. *The flood myth*. Berkeley: University of California Press.

Dunne, D., 2020. Is the climate crisis pushing the world towards a "point of no return"? *The Independent*, 13 November.

Durkheim, E., 2008 [1912]. *The elementary forms of the religious life*. Mineola, NY: Dover Publications.

Earp, B. D. & Trafimow, D. 2015. Replication, falsification, and the crisis of confidence in social psychology. *Frontiers in Psychology*, 6, p. 621.

Ecklund, E. H. & Scheitle, C. P., 2007. Religion among academic scientists: Distinctions, disciplines, and demographics. *Social Problems*, 54(2), pp. 289–307.

Ehrlich, P. R., 1968. *The population bomb*. New York: Ballantine Books.

Eliade, M., 1954. *The myth of the eternal return*. Princeton: Princeton University Press.

Elias, N., 1994. *The civilizing process*. Oxford: Blackwell.

Ellsberg, D., 2017. *The doomsday machine: Confessions of a nuclear war planner*. London: Bloomsbury.

Fairhead, J. & Leach, M., 2012. *Vaccine anxieties: Global science, child health and society*. London: Routledge.

Fava, S., 2013. *Environmental apocalypse in science and art: Designing nightmares*. London: Routledge.

Fawkes, G., 2019. *Greta Thunberg is daughter of Sweden's other Eurovision star* [Online] Available at: https://order-order.com/2019/04/24/greta-thunberg-privileged-daughter-swedens-eurovision-star/

Ferguson, M., 1980. *The aquarian conspiracy*. Los Angeles: J. P. Tarcher.

Festinger, L., 1962. *A theory of cognitive dissonance*. Redwood City, CA: Stanford University Press.

Festinger, L., Riecken, H. & Schachter, S., 1956. *When prophecy fails: A social and psychological study of a modern group that predicted the destruction of the world*. New York: Harper & Row.

Filiu, J.-P., 2011. *Apocalypse in Islam*. Berkeley: University of California Press.

Fitting, P., 2010. Utopia, dystopia and science fiction. In: G. Claeys, ed. *The Cambridge companion to utopian literature*. Cambridge: Cambridge University Press, pp. 135–153.

Fontenot, M. N., 2019. The art of eternal disaster: Tolkien's apocalypse and the road to healing. *Tolkien Studies* (16), pp. 91–109.

Forchtner, B., ed., 2019. *The far right and the environment: Politics, discourse and communication*. London: Routledge.

Fukuyama, F., 1992. *The end of history and the last man*. London: Penguin.

Fukuyama, F., 1995. Reflections on the end of history, five years later. *History and Theory*, 34(2), pp. 27–43.

Funtowicz, S. O. & Ravetz, J. R., 1990. *Uncertainty and quality in science for policy*. Dordrecht: Kluwer.

Funtowicz, S. O. & Ravetz, J. R., 1993. Science for the post-normal age. *Futures*, 25(7), pp. 739–755.

Furedi, F., 2006. *Culture of fear revisited*. London: Continuum.

Furseth, I. & Repstad, P., 2006. *An introduction to the sociology of religion: Classical and contemporary perspectives*. London: Routledge.

Garforth, L., 2005. Green utopias: Beyond apocalypse, progress, and pastoral. *Utopian Studies*, 16(1), pp. 1–37.

Garforth, L., 2018. *Green utopias: Environmental hope before and after nature*. Cambridge: Polity Press.

Garrett, B., 2020. *Bunker: Building for the end times*. London: Allan Lane.

Garrison, J., 1982. *The darkness of god: Theology after Hiroshima*. London: SCM Press.

Geoghegan, V., 2004. Ideology and utopia. *Journal of Political Ideologies*, 9(2), pp. 123–138.

Geraci, R. M., 2018. *Temples of modernity: Nationalism, Hinduism, and transhumanism in south Indian science*. Lanham: Rowman & Littlefield.

Giddens, A., 1991. *Modernity and self-identity: Self and society in the late modern age*. Stanford: Stanford University Press.

Giddens, A., 1999. Risk and responsibility. *The Modern Law Review*, 62, pp. 1–10.

Giddens, A., 2009. *Politics of climate change*. Cambridge: Polity Press.

Gieryn, T. F., 1999. *Cultural boundaries of science: Credibility on the line*. Chicago: University of Chicago Press.

Gigerenzer, G., 2004. Dread risk, September 11, and fatal traffic accidents. *Psychological Science*, 15(4), pp. 286–287.

Gillies, D. A., 1998. The Duhem thesis and the Quine thesis. In: M. Curd & J. Cover, eds. *Philosophy of science: The central issues*. New York: W. W. Norton, pp. 302–319.

Gillies, D. A., 2014. Should philosophers of mathematics make use of sociology? *Philosophia Mathematica*, 22(1), pp. 12–34.

Globus, R. & Taylor, B., 2011. Environmental millennialism. In: C. Wessinger, ed. *The Oxford handbook of millennialism*. Oxford: Oxford University Press, pp. 628–646.

Glock, C. Y. & Stark, R., 1965. *Religion and society in tension*. Chicago: Rand McNally.

Gnuse, R., 2011. Ancient near eastern millennialism. In: C. Wessinger, ed. *The Oxford handbook of millennialism*. Oxford: Oxford University Press, pp. 235–266.

Goffman, E., 1963. *Stigma: Notes on the management of a spoiled identity*. New York: Simon & Schuster.

Gonzalez, G. A., 2015. *The politics of Star Trek: Justice, war, and the future*. New York: Palgrave.

Gordin, M., Tilley, H. & Prakash, G., 2010. Introduction: Utopia and dystopia beyond space and time. In: M. Gordin, H. Tilley & G. Prakash, eds. *Utopia/dystopia: Conditions of historical possibility*. Princeton: Princeton University Press, pp. 1–17.

Gore, A., 1992. *Earth in the balance: Forging a new common purpose*. London: Routledge.

Gore, A., 2006. *An inconvenient truth: The planetary emergency of global warming and what we can do about it*. New York: Rodale.

Gould, S. J., 1997. Nonoverlapping magisteria. *Natural History*, 106, pp. 16–22.

Gray, J., 2007. *Black mass: Apocalyptic religion and the death of utopia*. London: Penguin.

Green, E., 2014. Half of Americans think climate change is a sign of the apocalypse. *The Atlantic*, 22 November.

Greenwood, J. J., 2007. Citizens, science and bird conservation. *Journal of Ornithology*, 148(1), pp. 77–124.

Gregory, J. & Miller, S., 1998. *Science in public*. London: Plenum.

Grundmann, R., 2012. The legacy of climategate: Revitalizing or undermining climate science and policy? *Wiley Interdisciplinary Reviews: Climate Change*, 3(3), pp. 281–288.

Guterres, A., 2020. 75 years after Hiroshima and Nagasaki, it's time to free the world of nuclear weapons. *The Independent*, 6 August.

Hacking, I., 2003. Risk and dirt. In: R. V. Ericson & A. Doyle, eds. *Risk and morality*. Toronto: University of Toronto Press, pp. 22–47.

Hales, P. B., 1991. The atomic sublime. *American Studies*, 32(1), pp. 5–31.

Hall, J. R., 2009. *Apocalypse: From antiquity to the empire of modernity*. Cambridge: Polity Press.

Hamblin, J. D., 2013. *Arming mother nature: The birth of catastrophic environmentalism*. Oxford: Oxford University Press.

Hamilton, A., 2020. Conservatism. In: E. N. Zalta, ed. *The Stanford encyclopedia of philosophy*. Spring 2020 Edition. [Online] Available at: https://plato.stanford.edu/archives/spr2020/entries/conservatism/ (accessed 21.01.2021).

Hanegraaff, W. J., 1999. New age spiritualities as secular religion: A historian's perspective. *Social Compass*, 46(2), pp. 145–160.

Hanegraaff, W. J., 2007. The new age movement and Western esotericism. In: D. Kemp & J. R. Lewis, eds. *Handbook of new age*. Leiden: Brill, pp. 25–50.

Hanegraaff, W. J., 2013. *Western esotericism: A guide for the perplexed*. London: Bloomsbury.

Harambam, J., 2020. *Contemporary conspiracy Culture: Truth and knowledge in an era of epistemic instability*. London: Routledge.

Harambam, J. & Aupers, S., 2015. Contesting epistemic authority: Conspiracy theories on the boundaries of science. *Public Understanding of Science*, 24(4), pp. 466–480.

Harambam, J. & Aupers, S., 2017. "I am not a conspiracy theorist": Relational identifications in the Dutch conspiracy milieu. *Cultural Sociology*, 11(1), pp. 113–129.

Haraway, D., 2015. Anthropocene, capitalocene, plantationocene, chthulucene: Making kin. *Environmental Humanities*, 6(1), pp. 159–165.

Harrison, P., ed., 2010. *The Cambridge companion to science and religion*. Cambridge: Cambridge University Press.

Harrison, P., 2015. *The territories of science and religion*. Chicago: University of Chicago Press.

Hassler-Forest, D., 2017. Mad Max: Between apocalypse and utopia. *Science Fiction Film and Television*, 10(3), pp. 301–306.

Hawkins, E. & Sutton, R., 2012. Time of emergence of climate signals. *Geophysical Research Letters*, 39(1).

Hersey, J., 1946. Hiroshima. *The New Yorker*, 24 August.

Hill, M. O., 1994. *Dreaming the end of the world: Apocalypse as a rite of passage*. Dallas: Spring Publications.

Hofstadter, R., 1966. *The paranoid style in American politics and other essays*. New York: Knopf.

Hogg, M. A. & Abrams, D. E., 1988. *Social identifications*. London: Routledge.

Höijer, B., 2010. Emotional anchoring and objectification in the media reporting on climate change. *Public Understanding of Science*, 19(6), pp. 717–731.

Horn, E., 2014. *Zukunft als Katastrophe*. Frankfurt am Main: Fischer Verlag.

Hughes, J. A., 2003. *The Manhattan project: Big science and the atom bomb*. Cambridge: Icon Books.

Hulme, M., 2009. *Why we disagree about climate change: Understanding controversy, inaction and opportunity*. Cambridge: Cambridge University Press.

Huxley, A., 1932. *Brave new world*. London: Chatto and Windus.

Irwin, A., 2001. *Sociology and the environment*. Cambridge: Polity Press.

James, L., 1993. From Robinson to Robina, and beyond: Robinson Crusoe as a utopian concept. In: K. Kumar & S. Bann, eds. *Utopias and the millennium*. London: Reaktion Books.

Jameson, F., 2005. *Archaeologies of the future: The desire called utopia and other science fictions*. London: Verso.

Jamison, A., 2001. *The making of green knowledge: Environmental politics and cultural transformation*. Cambridge: Cambridge University Press.

Jaspers, K., 1961. *Die Atombombe und die Zukunft des Menschen*. München: Deutscher Taschenbuch Verlag.

Jenkins, T., 2013. *Of flying saucers and social scientists*. New York: Palgrave.

Joffe, H., 1999. *Risk and the other*. Cambridge: Cambridge University Press.

Jong, J., 2015. On (not) defining (non) religion. *Science, Religion and Culture*, 2(3), pp. 15–24.

Jovchelovitch, S., 2007. *Knowledge in context: Representations, community and culture*. London: Routledge.

Jovchelovitch, S. & Hawlina, H., 2018. Utopias and world-making: Time, transformation and the collective imagination. In: C. D. Saint-Laurent, S. Obradovic & K. Carriere, eds. *Imagining collective futures*. London: Palgrave Macmillan, pp. 129–151.

Juergensmeyer, M., 2017. *Terror in the mind of God: The global rise of religious violence*. Berkeley: University of California Press.

Kahan, D. M., 2012. Why we are poles apart on climate change. *Nature*, 488(7411), p. 255.

Kahan, D. M., Jenkins-Smith, H. & Braman, D., 2011. Cultural cognition of scientific consensus. *Journal of Risk Research*, 14(2), pp. 147–174.

Karplus, W. J., 1992. *The heavens are falling: The scientific prediction of catastrophes in our time*. New York: Plenum.

Kashima, Y. & Fernando, J., 2020. Utopia and ideology in cultural dynamics. *Current Opinion in Behavioral Sciences*, 34, pp. 102–106.

Kermode, F., 1967. *The sense of an ending: Studies in the theory of fiction*. Oxford: Oxford University Press.

Kettler, D., Meja, V. & Stehr, N., 1990. Rationalizing the irrational: Karl Mannheim and the besetting sin of German intellectuals. *American Journal of Sociology*, 95(6), pp. 1441–1473.

Khan, I., 2008. The apocalypse has been postponed: We apologise for any inconvenience. *The Guardian*, 25 September.

Knowles, S., 2018. Brexit, Babylon and prophecy: Semiotics of the end times. *Religions*, 9(12), p. 396.

Koch, I., 2017. What's in a vote? Brexit beyond culture wars. *American Ethnologist*, 44(2), pp. 225–230.

Kreyenbroek, P. G., 2002. Millennialism and eschatology in the Zoroastrian tradition. In: A. Amanat & M. T. Bernhardsson, eds. *Imagining the end: Visions of apocalypse from the ancient Middle East to modern America*. London: I. B. Tauris, pp. 33–55.

Kuhn, T., 1962. *The structure of scientific revolutions*. Chicago: University of Chicago Press.

Kumar, K., 1991. *Utopianism*. Milton Keynes: Open University Press.

Kumar, K., 2006. Ideology and sociology: Reflections on Karl Mannheim's ideology and utopia. *Journal of Political Ideologies*, 11(2), pp. 169–181.

Lahr, A. M., 2007. *Millennial dreams and apocalyptic nightmares*. Oxford: Oxford University Press.

Lakatos, I., 1978. *Philosophical papers volume I: The methodology of scientific research programmes*. Cambridge: Cambridge University Press.

Lakatos, I. & Zahar, E. G., 1976. Why did Copernicus's programme supersede Ptolemy's? In: R. Westman, ed. *The Copernican achievement*. Los Angeles: University of California Press, pp. 354–383.

Lamont, M. & Molnár, V., 2002. The study of boundaries in the social sciences. *Annual Review of Sociology*, 28(1), pp. 167–195.

Landes, R., 2011. *Heaven on earth: The varieties of the millennial experience*. Oxford: Oxford University Press.

Latour, B., 2017. *Facing Gaia: Eight lectures on the new climatic regime*. Cambridge: Polity Press.

Lawrence, D., 2020. *The UK's emerging conspiracy theory street movements*. [Online] Available at: www.hopenothate.org.uk/2020/08/28/the-uks-emerging-conspiracy-theory-street-movements/

Leithart, P. J., 2018. *Revelation 1–11*. London: Bloomsbury T&T Clark.

Lenton, T. M. et al., 2019. Climate tipping points: Too risky to bet against. *Nature*, 575, pp. 592–595.

Lesley, J., 1996. *The end of the world: The science and ethics of human extinction*. London: Routledge.

Letwin, O., 2020. *Apocalypse how*. London: Atlantic Books.

Levinovitz, A., 2020. *Natural: The seductive myth of nature's goodness*. London: Profile Books.

Levitas, R., 1990. *The concept of utopia*. Oxford: Peter Lang.

Lewis, S. L. & Maslin, M. A., 2015. Defining the Anthropocene. *Nature*, 519(7542), pp. 171–180.

Lindberg, D. C., 2010. The fate of science in patristic and medieval Christendom. In: *The Cambridge companion to science and religion*. Cambridge: Cambridge University Press, pp. 21–38.

Lindsey, H., 1970. *The late great planet earth*. Grand Rapids: Zondervan.

Lindsey, H., 1994. *Planet earth: 2000 AD: Will mankind survive?* Palos Verdes, CA: Western Front.

Lovelock, J., 1979. *Gaia: A new look at life on earth*. Oxford: Oxford University Press.

Lowe, S., 2000. Western millennial ideology goes east: The Taiping rebellion and Mao's great leap forward. In: C. Wessinger, ed. *Millennialism, persecution, and violence: Historical cases*. New York: Syracuse University Press, pp. 220–240.

Lupton, D., 1999. *Risk*. London: Routledge.

Malthus, T., 1992 [1798]. *Essay on the principle of population*. Cambridge: Cambridge University Press.

Mann, M. E., 2014. *The hockey stick and the climate wars: Dispatches from the front lines*. New York: Columbia University Press.

Mannheim, K., 2015 [1936]. *Ideology and utopia*. Mansfield Centre, CT: Martino Publishing.

Manzo, K., 2010. Beyond polar bears? Re-envisioning climate change. *Meteorological Applications*, 17(2), pp. 196–208.

Markham, W. T., 2008. *Environmental organizations in modern Germany: Hardy survivors in the twentieth century and beyond*. New York: Berghahn Books.

Marx, K., 1978. *The Marx-Engels reader*. New York: W. W. Norton.

Mason, P., 2016. Are we living through another 1930s? *The Guardian*, 1 August.

McDonagh, E. L., 1976. Attitude changes and paradigm shifts: Social psychological foundations of the Kuhnian thesis. *Social Studies of Science*, 6(1), pp. 51–76.

McGhee, G. S., 2005. A cultural history of dissonance theory. In: S. O'Leary & G. McGhee, eds. *War in heaven, heaven on earth: Theories of the apocalyptic*. London: Routledge, pp. 195–220.

McKeon, M., 1987. *The origins of the English novel, 1600–1740*. Baltimore: Johns Hopkins University Press.

McMinn, L., 2001. Y2K, the apocalypse, and evangelical Christianity: The role of eschatological belief in church responses. *Sociology of Religion*, 62(2), pp. 205–220.

McNeish, W., 2017. From revelation to revolution: Apocalypticism in green politics. *Environmental Politics*, 26(6), pp. 1035–1054.

Meadows, D., Meadows, D., Randers, J. & Behrens, W., 1972. *Limits to growth*. New York: Universe Books.

Mellor, P. A., 1998. Sacred contagion and social vitality: Collective effervescence in "Les Formes élémentaires de la vie religieuse". *Durkheimian Studies/Etudes Durkheimiennes*, 4, pp. 87–114.

Merton, R. K., 1973. *The sociology of science: Theoretical and empirical investigations*. Chicago: University of Chicago Press.

Miller, J. D., 1998. The measurement of civic scientific literacy. *Public Understanding of Science*, 7(3), pp. 203–224.

Miller, W. M., 1993 [1959]. *A canticle for Leibowitz*. London: Orbit.

Moore, W. E., 1966. The utility of utopias. *American Sociological Review*, 31(6), pp. 765–772.

More, T., 1989 [1516]. *Utopia*. Cambridge: Cambridge University Press.

Moritz, J. M., 2009. Rendering unto science and god: Is NOMA enough? *Theology and Science*, 7(4), pp. 363–378.

Morton, T., 2013. *Hyperobjects: Philosophy and ecology after the end of the world*. Cambridge: Harvard University Press.

Moscovici, S., 1981. On social representations. *Social Cognition: Perspectives on Everyday Understanding*, 8(12), pp. 181–209.

Moscovici, S., 2001. *Social representations: Essays in social psychology*. New York: New York University Press.

Moscovici, S., 2008. *Psychoanalysis: Its image and its public*. Cambridge: Polity Press.

Moses, A. D., 2019. "White genocide" and the ethics of public analysis. *Journal of Genocide Research*, 21(2), pp. 201–213.

Mueller, J., 2010. *Atomic obsession: Nuclear alarmism from Hiroshima to Al-Qaeda*. Oxford: Oxford University Press.

Musgrave, A. & Pigden, C., 2016. Imre Lakatos. In: E. N. Zalta, ed. *Stanford encyclopedia of philosophy*. [Online] Available at: https://plato.stanford.edu/archives/win2016/entries/lakatos/ (accessed 21.01.2021).

Nagle, A., 2017. *Kill all normies: Online culture wars from 4chan and Tumblr to Trump and the alt-right*. Alresford: Zero.

Neuneck, G., 2014. Carl Friedrich von Weizsäcker: Nukleare Abrüstung und die Suche nach Frieden. In: D. H. Klaus Hentschel, ed. *Carl Friedrich von Weizsäcker: Physik-Philosophie-Friedensforschung*. Stuttgart: Wissenschaftliche Verlagsgesellschaft, pp. 413–436.

Nongbri, B., 2013. *Before religion*. New Haven, CT: Yale University Press.

Nordhaus, T. & Shellenberger, M., 2009. Apocalypse fatigue: Losing the public on climate change. *Yale Environment 360*, 16 November.

O'Leary, S. D., 1994. *Arguing the apocalypse: A theory of millennial rhetoric*. Oxford: Oxford University Press.

O'Neill, B., 2019. The cult of Greta Thunberg. *Spiked Online*, 22 April.

Ord, T., 2020. *The precipice: Existential risk and the future of humanity*. London: Bloomsbury.

Oreskes, N. & Conway, E. M., 2011. *Merchants of doubt: How a handful of scientists obscured the truth on issues from tobacco smoke to global warming*. London: Bloomsbury.

Orwell, G., 1946. *Animal farm*. New York: Harcourt.

Orwell, G., 1949. *Nineteen eighty-four*. New York: Harcourt.

Osborne, G. R., 2002. *Revelation verse by verse*. Grand Rapids, MI: Baker Academic.

Paskins, M., 2020. History of science and its utopian reconstructions. *Studies in History and Philosophy of Science, Part A*, 81, pp. 82–95.

Pepper, D., 1996. *Modern environmentalism: An introduction*. London: Routledge.

Pesantubbee, M. E., 2000. From vision to violence: The Wounded Knee massacre. In: C. Wessinger, ed. *Millennialism, persecution, & violence*. New York: Syracuse University Press, pp. 62–81.

Peyser, T. G., 1992. Reproducing utopia: Charlotte Perkins Gilman and Herland. *Studies in American Fiction*, 20(1), pp. 1–16.

Pike, N., 1964. Hume on evil. In: N. Pike, ed. *God and evil*. Englewood Cliffs: Prentice-Hall, pp. 85–102.

Polanyi, M., 1962. The republic of science. *Minerva*, 1(1), pp. 54–73.

Popper, K., 1963. *Conjectures and refutations*. London: Routledge.

Popper, K., 2005 [1959]. *The logic of scientific discovery*. London: Routledge.

Popper, K., 2011 [1945]. *The open society and its enemies*. London: Routledge.

Poushter, J. & Huang, C. 2019. *Climate change still seen as the top global threat, but cyberattacks a rising concern*. Pew Research Center. [Online] Available at: https://www.pewresearch.org/global/wp-content/uploads/sites/2/2019/02/Pew-Research-Center_Global-Threats-2018-Report_2019-02-10.pdf

Prainsack, B. & Riesch, H., 2016. Interdisciplinarity reloaded? Drawing lessons from "citizen science". In: S. Frickel, M. Albert & B. Prainsack, eds. *Investigating interdisciplinary collaboration: Theory and practice across disciplines*. New Brunswick, NJ: Rutgers University Press, pp. 194–212.

Prats, J., 2012. The L'Aquila earthquake: Science or risk on trial? *Significance*, 9(6), pp. 13–16.

Quine, W.V.O., 1980 [1953]. *From a logical point of view*. Cambridge, MA: Harvard University Press.

Rapp, C., 2010. Aristotle's rhetoric. In: E. N. Zalta, ed. *The Stanford encyclopedia of philosophy*. Spring 2010 Edition. [Online] Available at: https://plato.stanford.edu/archives/spr2010/entries/aristotle-rhetoric

Redles, D., 2011. National socialist millennium. In: C. Wessinger, ed. *The Oxford handbook of millennialism*. Oxford: Oxford University Press, pp. 529–548.

Rees, M., 2004. *Our final century*. London: Arrow books.

Reeves, M., 1980. The originality and influence of Joachim of Fiore. *Traditio*, 36, pp. 269–316.

Relf, J., 1993. Utopia the good breast: Coming home to mother. In: K. Kumar & S. Bann, eds. *Utopias and the millennium*. London: Reaktion Books, pp. 107–128.

Rieh, S. Y. & Danielson, D. R., 2007. Credibility: A multidisciplinary framework. In: B. Cronin, ed. *Annual review of information science and technology*. Vol. 41. Medford: Information Today, pp. 307–364.

Riesch, H., 2008. *Scientists' views of the philosophy of science*. Unpublished PhD thesis. London: University College London.

Riesch, H., 2010. Theorizing boundary work as representation and identity. *Journal for the Theory of Social Behaviour*, 40(4), pp. 452–473.

Riesch, H., 2013. Levels of uncertainty. In: S. Roeser, R. Hillerbrand, P. Sandin & M. Peterson, eds. *Essentials of risk theory*. Dordrecht: Springer, pp. 29–56.

Riesch, H., 2014. Philosophy, history and sociology of science: Interdisciplinary relations and complex social identities. *Studies in History and Philosophy of Science Part A*, 48, pp. 30–37.

Riesch, H., forthcoming. Citizen science, environmental crisis and redemption. *Studia Sociologica: Annales Universitatis Paedagogicae Cracoviensis*.

Riesch, H. et al., 2013. Internet-based public debate of CCS: Lessons from online focus groups in Poland and Spain. *Energy Policy*, 56, pp. 693–702.

Robertson, D. G., 2016. *UFOs, conspiracy theories and the new age: Millennial conspiracism*. London: Bloomsbury.

Robinson, K. S., 2003. *The years of rice and salt*. New York: Harper Collins.

Rogińska, M., 2019. Trajectories of (non)belief in the scientific community: The case of Polish and Ukrainian natural scientists. *Review of Religious Research*, 61, pp. 389–409.

Rohkrämer, T., 2002. Contemporary environmentalism and its links with the German past. In: *The cultures of German environmentalism*. New York: Berghahn, pp. 48–62.

Rohloff, A., 2012. *Climate change, moral panic, and civilization: On the development of global warming as a social problem*. Unpublished PhD dissertation. London: Brunel University London.

Rohloff, A., 2019. *Climate change, moral panics, and civilization*. London: Routledge.

Rose, H. & Rose, S., 1979. Radical science and its enemies. *Socialist Register*, 16, pp. 317–335.

Rosenfeld, J. A., 2011. Nativist millennialism. In: C. Wessinger, ed. *The Oxford handbook of millennialism*. Oxford: Oxford University Press, pp. 89–112.

Russell, B., 1957. *Why I am not a Christian: And other essays on religion and related subjects*. New York: Simon & Schuster.

Russell, B., 2010 [1959]. *Common sense and nuclear warfare*. London: Routledge.

Saler, B., 1999. Family resemblance and the definition of religion. *Historical Reflections/ Reflexions historiques*, 25(3), pp. 391–404.

Sammut, G., Andreouli, E., Gaskell, G. & Valsiner, J., 2015. Social representations: A revolutionary paradigm? In: G. Sammut, E. Andreouli, G. Gaskell & J. Valsiner, eds. *Cambridge handbook of social representations*. Cambridge: Cambridge University Press, pp. 3–11.

Sánchez-Bayo, F. & Wyckhuys, K. A., 2019. Worldwide decline of the entomofauna: A review of its drivers. *Biological Conservation*, 232, pp. 8–27.

Schäfer, W., 2013. Plutoniumbombe und zivile Atomkraft: Carl Friedrich von Weizsäckers Beiträge zum Dritten Reich und zur Bundesrepublik. *Leviathan*, 41(3), pp. 383–421.

Schell, J., 1982. *The fate of the earth*. New York: Alfred A. Knopf.

Scheller, J., 2020. Lob auf die Apokalypse. *Neue Zürcher Zeitung*, 7 April.

Schmalz, M. N., 1994. When Festinger fails: Prophecy and the watchtower. *Religion*, 24(4), pp. 293–308.

Shapin, S., 1996. *The scientific revolution*. Chicago: University of Chicago Press.

Shapiro, J. F., 2002. *Atomic bomb cinema: The apocalyptic imagination on film*. London: Routledge.

Shils, E., 1974. Ideology and utopia by Karl Mannheim. *Daedalus*, 103(1), pp. 83–89.

Sideris, L. H. & Whalen-Bridge, J., 2019. Special issue introduction: Popular culture, religion, and the Anthropocene. *Journal for the Study of Religion, Nature and Culture*, 13(4), pp. 409–413.

Skrimshire, S., 2008. *Politics of fear, practices of hope: Depoliticisation and resistance in a time of terror.* London: Continuum.

Skrimshire, S., 2010. Eternal return of apocalypse. In: S. Skrimshire, ed. *Future ethics: Climate change and apocalyptic imagination.* London: Continuum, pp. 219–241.

Skrimshire, S., 2014. Climate change and apocalyptic faith. *Wiley Interdisciplinary Reviews: Climate Change,* 5(2), pp. 233–246.

Slovic, P., 1987. Perception of risk. *Science,* 236(4799), pp. 280–285.

Smart, N., 1989. *The world's religions.* Cambridge: Cambridge University Press.

Smith, P. & Howe, N., 2015. *Climate change as social drama: Global warming in the public sphere.* Cambridge: Cambridge University Press.

Snowden, F. M., 2019. *Epidemics and society: From the black death to the present.* New Haven: Yale University Press.

Sonne, W., 2018. *Leben mit der Bombe.* Wiesbaden: Springer.

Spiegelhalter, D. J. & Riesch, H., 2011. Don't know, can't know: Embracing deeper uncertainties when analysing risks. *Philosophical Transactions of the Royal Society A: Mathematical, Physical and Engineering Sciences,* 369(1956), pp. 4730–4750.

Spoel, P., Goforth, D., Cheu, H. & Pearson, D., 2008. Public communication of climate change science: Engaging citizens through apocalyptic narrative explanation. *Technical Communication Quarterly,* 18(1), pp. 49–81.

Star, S. L. & Griesemer, J. R., 1989. Institutional ecology, translations' and boundary objects: Amateurs and professionals in Berkeley's museum of Vertebrate Zoology, 1907–39. *Social Studies of Science,* 19(3), pp. 387–420.

Stark, R., 1964. Class, radicalism, and religious involvement in Great Britain. *American Sociological Review,* 29(5), pp. 698–706.

Steffen, W., Grinevald, J., Crutzen, P. & McNeill, J., 2011. The Anthropocene: Conceptual and historical perspectives. *Philosophical Transactions of the Royal Society A: Mathematical, Physical and Engineering Sciences,* 369(1938), pp. 842–867.

Stenmark, M., 2010. Ways of relating science and religion. In: P. Harrison, ed. *The Cambridge companion to science and religion.* Cambridge: Cambridge University Press, pp. 278–295.

Stern, N. H., 2007. *The economics of climate change: The Stern review.* Cambridge: Cambridge University Press.

Steyn, C., 2000. Millenarian tragedies in South Africa: The Xhosa cattle-killing movement and the Bulhoek massacre. In: C. Wessinger, ed. *Millennialism, persecution & violence.* New York: Syracuse University Press, pp. 185–204.

Stirling, A., 2007. Risk, precaution and science: Towards a more constructive policy debate: Talking point on the precautionary principle. *EMBO Reports,* 8(4), pp. 309–315.

Stoll, M. R., 2001. Green versus green: Religions, ethics, and the Bookchin-Foreman dispute. *Environmental History,* 6(3), pp. 412–427.

Stoll, M. R., 2015. *Inherit the holy mountain: Religion and the rise of American environmentalism.* Oxford: Oxford University Press.

Stone, J. R. ed., 2000. *Expecting Armageddon: Essential reading in failed prophecy.* London: Routledge.

Sutton, P. W., 2007. *The environment: A sociological introduction.* Cambridge: Polity Press.

Swyngedouw, E., 2010. Apocalypse forever? *Theory, Culture & Society,* 27(2–3), pp. 213–232.

Swyngedouw, E., 2013. Apocalypse now! Fear and doomsday pleasures. *Capitalism Nature Socialism,* 24(1), pp. 9–18.

Tabor, J. D., 2011. Ancient Jewish and early Christian millennialism. In: C. Wessinger, ed. *The Oxford handbook of millennialism*. Oxford: Oxford University Press, pp. 252–283.

Tajfel, H., 1974. Social identity and intergroup behaviour. *Information (International Social Science Council)*, 13(2), pp. 65–93.

Tajfel, H., 1981. *Human groups and social categories: Studies in social psychology*. Cambridge: Cambridge University Press.

Tajfel, H., 1982. Social psychology of intergroup relations. *Annual Review of Psychology*, 33, pp. 1–39.

Tajfel, H. & Wilkes, A. L., 1963. Classification and quantitative judgement. *British Journal of Social Psychology*, 54(2), pp. 101–114.

Tamimi, N., 2018. The ambiguity of e-cigarettes: E-cigarettes as boundary objects. *Journal of Integrated Social Sciences*, 8(1), pp. 68–88.

Taylor, B., 2010. *Dark green religion: Nature spirituality and the planetary future*. Berkeley: University of California Press.

Taylor, B., 2016. The greening of religion hypothesis (part one): From Lynn White, Jr and claims that religions can promote environmentally destructive attitudes and behaviors to assertions they are becoming environmentally friendly. *Journal for the Study of Religion, Nature and Culture*, 10(3), pp. 268–305.

Taylor, B., 2019. Alt-right ecology: Ecofascism and far-right environmentalism in the United States. In: B. Forchtner, ed. *The far right and the environment: Politics, discourse and communication*. London: Routledge, pp. 275–292.

Taylor, C., 2009. *A secular age*. Cambridge, MA: Harvard University Press.

Thornhill, C. & Miron, R., 2020. Karl Jaspers. In: E. N. Zalta, ed. *The Stanford encyclopedia of philosophy*. Spring 2020 Edition. [Online] Available at: https://plato.stanford.edu/archives/spr2020/entries/jaspers/

Thorpe, C., 2008. *Oppenheimer: The tragic intellect*. Chicago: Chicago University Press.

Thunberg, G., 2019. *No one is too small to make a difference*. London: Penguin.

Treloar, J. L., 1988. Tolkien and Christian concepts of evil: Apocalypse and privation. *Mythlore*, 15(2 [56]), pp. 57–60.

Trompf, G. W., 2011. Pacific millennial movements. In: C. Wessinger, ed. *The Oxford handbook of millennialism*. Oxford: Oxford University Press, pp. 435–456.

Turney, J., 1998. *Frankenstein's footsteps: Science, genetics and popular culture*. New Haven: Yale University Press.

Turney, J., 2003. *Gaia: Signs of life*. Cambridge: Icon Books.

United Nations, 2019. *World population prospects 2019 highlights*. [Online] Available at: https://population.un.org/wpp/Publications/Files/WPP2019_Highlights.pdf

Universität Göttingen, n.d. *The manifesto*. [Online] Available at: www.uni-goettingen.de/en/the+manifesto/54320.html

University of Reading, 2011. *Climate change could affect tropics in ten years*. [Online] Available at: www.reading.ac.uk/news-archive/press-releases/pr423941.html

van Fossen, A. B., 1988. How do movements survive failures of prophecy. *Research in Social Movements, Conflicts and Change*, 10, pp. 193–212.

von Stuckrad, K., 2014. *The scientification of religion: An historical study of discursive change, 1800–2000*. Boston: de Gruyter.

Wackernagel, M. & Rees, W., 1996. *Our ecological footprint: Reducing human impact on the earth*. Gabriola Island, BC: New Society Publishers.

Wagar, W. W., 1982. *Terminal visions: The literature of last things*. Bloomington: Indiana University Press.

Waldzus, S., Mummendey, A., Wenzel, M. & Boettcher, F., 2004. Of bikers, teachers and Germans: Groups' diverging views about their prototypicality. *British Journal of Social Psychology*, 43(3), pp. 385–400.

Walliss, J. & Newport, K., eds., 2014. *The end all around us: Apocalyptic texts and popular culture*. London: Routledge.

Walsh, L., 2013. *Scientists as prophets: A rhetorical genealogy*. Oxford: Oxford University Press.

Washer, P., 2004. Representations of SARS in the British newspapers. *Social Science & Medicine*, 59(12), pp. 2561–2571.

Watt, C. S., 2020. The QAnon orphans: People who have lost loved ones to conspiracy theories. *The Guardian*, 23 September.

Watts, P. M., 1985. Prophecy and discovery: On the spiritual origins of Christopher Columbus's "Enterprise of the Indies". *The American Historical Review*, 90(1), pp. 73–102.

Weart, S. R., 1988. *Nuclear fear: A history of images*. Cambridge, MA: Harvard University Press.

Weart, S. R., 2008. *The discovery of global warming*. New York: New York University Press.

Webb, J. & Byrnand, S., 2008. Some kind of virus: The zombie as body and as trope. *Body & Society*, 14(2), pp. 83–98.

Weber, E., 1999. *Apocalypses: Prophecies, cults, and millennial beliefs through the ages*. Cambridge, MA: Harvard University Press.

Weber, M., 1963 [1922]. *The sociology of religion*. London: Methuen & Co. Ltd.

Weber, M., 1992 [1930]. *The Protestant ethic and the spirit of capitalism*. London: Routledge.

Weingart, P., Engels, A. & Pansegrau, P., 2000. Risks of communication: Discourses on climate change in science, politics, and the mass media. *Public Understanding of Science*, 9(3), pp. 261–284.

Wells, H. G., 1914. *The world set free: A story of mankind*. New York: Dutton.

Wessinger, C., ed., 2000. *Millennialism, persecution, & violence: Historical cases*. New York: Syracuse University Press.

Wessinger, C., 2011a. Millennial glossary. In: C. Wessinger, ed. *The Oxford handbook of millennialism*. Oxford: Oxford University Press, pp. 717–724.

Wessinger, C., ed., 2011b. *The Oxford handbook of millennialism*. Oxford: Oxford University Press.

White, L., 1967. The historical roots of our ecologic crisis. *Science*, 155(3767), pp. 1203–1207.

Whitney, E., 2015. Lynn White Jr.'s "the historical roots of our ecologic crisis" after 50 years. *History Compass*, 13(8), pp. 396–410.

Williams, R., 1978. Utopia and science fiction. *Science Fiction Studies*, 5(3), pp. 203–214.

Wirth, L., 2015 [1936]. Preface. In: *Ideology and utopia*. Mansfield Centre, CT: Martino Publishing, pp. xi–xxxi.

Wittgenstein, L., 1953. *Philosophical investigations*. London: Blackwell.

Wojcik, D., 1996. Embracing doomsday: Faith, fatalism, and apocalyptic beliefs in the nuclear age. *Western Folklore*, 55(4), pp. 297–330.

Wojcik, D., 1997. *The end of the world as we know it: Faith, fatalism and apocalypse in America*. New York: New York University Press.

Wojcik, D., 2011. Avertive apocalypticism. In: C. Wessinger, ed. *The Oxford handbook of millennialism*. Oxford: Oxford University Press, pp. 66–88.

World Commission on Environment and Development, 1987. *Our common future*. Oxford: Oxford University Press.

Worrall, J., 2004. Does science discredit religion? In: M. Peterson & R. Vanarragon, eds. *Contemporary debates in philosophy of religion*. Malden, MA: Blackwell, pp. 58–72.

Wuthnow, R., 2010. *Be very afraid: The cultural response to terror, pandemics, environmental devastation, nuclear annihilation, and other threats*. Oxford: Oxford University Press.

Yearley, S., 1992a. *The green case: A sociology of environmental issues, arguments and politics*. London: Routledge.

Yearley, S., 1992b. Green ambivalence about science: Legal-rational authority and the scientific legitimation of a social movement. *British Journal of Sociology*, 43(4), pp. 511–532.

Yearley, S., 2004. *Making sense of science: Understanding the social study of science*. London: Sage.

Ziemann, B., 2016. German angst? Debating cold war anxieties in West Germany, 1945–90. In: M. Grant & B. Ziemann, eds. *Understanding the imaginary war*. Manchester: Manchester University Press, pp. 116–139.

Žižek, S., 2010. *Living in the end times*. London: Verso.

Index

For Product Safety Concerns and Information please contact our EU
representative GPSR@taylorandfrancis.com
Taylor & Francis Verlag GmbH, Kaufingerstraße 24, 80331 München, Germany

www.ingramcontent.com/pod-product-compliance
Lightning Source LLC
Chambersburg PA
CBHW071113100726
47908CB00008B/2362